Reproductive Behavior and Evolution

EVOLUTION, DEVELOPMENT, AND ORGANIZATION OF BEHAVIOR

Daniel S. Lehrman Memorial Symposia Series

Series Editors: Jay S. Rosenblatt and B. R. Komisaruk
Rutgers University

VOLUME 1 • REPRODUCTIVE BEHAVIOR AND EVOLUTION
Edited by Jay S. Rosenblatt and B.R. Komisaruk

A Continuation Order Plan is available for this series. A continuation order will bring delivery of each new volume immediately upon publication. Volumes are billed only upon actual shipment. For further information please contact the publisher.

Reproductive Behavior and Evolution

Edited by

Jay S. Rosenblatt

and

B. R. Komisaruk

Rutgers University
Newark, New Jersey

Plenum Press · New York and London

Library of Congress Cataloging in Publication Data

Main entry under title:

Reproductive behavior and evolution.

(Evolution, development, and organization of behavior; v. 1)
Includes bibliographies.
1. Sexual behavior in animals. 2. Evolution. I. Rosenblatt, Jay S. II.
Komisaruk, B. R. III. Series.
QL761.R46   591.5'6   77-10855
ISBN 0-306-34481-5

© 1977 Plenum Press, New York
A Division of Plenum Publishing Corporation
227 West 17th Street, New York, N.Y. 10011

Printed in the United States of America

Contributors

J. H. Crook, Department of Psychology, University of Bristol, United Kingdom

Julian M. Davidson, Department of Physiology, Stanford University, Stanford, California 94305

John F. Eisenberg, National Zoological Park, Smithsonian Institution, Washington, D.C.

Paul D. MacLean, Laboratory of Brain Evolution and Behavior, National Institute of Mental Health, Bethesda, Maryland 20014

Ernst Mayr, Museum of Comparative Zoology, The Agassiz Museum, Harvard University, Cambridge, Massachusetts 02138

Thomas E. McGill, Department of Psychology, Williams College, Williamstown, Massachusetts 01267

Berta Scharrer, Department of Anatomy, Albert Einstein College of Medicine, Bronx, New York 10461

Preface

The Daniel S. Lehrman Memorial Symposia Series will publish the proceedings of symposia devoted to the evolution, development, and organization of behavior. These various symposia will bring together at intervals scientists studying problems from each of these viewpoints. The aim is to attempt to integrate our knowledge derived from these different sources and to familiarize scientists working on similar behavior patterns with the work of their colleagues in related fields of study. Each volume, therefore, will be devoted to a specific topic in the field of animal behavior, which will be explored with respect to its evolutionary aspects, including the adaptive nature of the behavior; with respect to its developmental aspects, including neural, hormonal, and experiential influences; and with respect to the analysis of features of organization, including motivational, perceptual, and motor aspects and their physiological bases. It is our feeling that the most appropriate memorial to our colleague and close friend, Daniel S. Lehrman, is the continuation of his valuable contributions toward integrating these approaches to the study of animal behavior, which he pursued so effectively during his own life.

Daniel S. Lehrman's lifelong love and study of animal behavior gave us a wealth of new insights into reproductive behavior and evolution. It is therefore appropriate that the first symposium of this series is devoted to recent advances in this field. This topic was chosen for our first symposium also because of what we believe is the main purpose of these symposia: to stimulate discussion among investigators with different methods and approaches to a common problem.

Reproductive behavior and evolution is a particularly suitable topic for this kind of discussion as the articles in the volume will show. An evolutionary biologist, Ernst Mayr, discusses the concepts of *innate* and *learned* from an evolutionary point of view as they apply to reproductive and nonreproductive behavior. John H. Crook, a psychologist, and John Eisenberg, an ethologist, provide an evolutionary perspective on reproductive behavior in relation to the variety of social forms of adaptation to different ecological conditions among mammals. Implications about reproductive behavior as the product of natural selection and as a reproductive isolating mechanism in evolution are discussed by Thomas E. McGill, a behavioral geneticist. Neuroendocrine mechanisms underlying reproductive behavior, which mediate adaptation to external stimuli, are discussed chiefly among invertebrates by Berta Scharrer, a pioneer neuroendocrinologist, and among mammals by Julian Davidson, who has contributed richly to our understanding of mammalian neuroendocrine mechanisms. Paul MacLean, a neuropsychiatrist, discusses the evolution of neural structures mediating complex social interactions involved in reproductive behavior.

We believe that research scientists studying reproductive behavior and evolution are more likely to make progress in their own field and to contribute to closely related fields if they are aware of developments in these fields and how their own research can contribute to, as well as benefit from, these developments.

Jay S. Rosenblatt
B. R. Komisaruk

Newark

Contents

Chapter 4

Chapter 5

Concepts in the Study of Animal Behavior

Ernst Mayr

Several of Danny Lehrman's papers show what a great interest he had in the nature of the concepts that are used by students of animal behavior. Historians of science have demonstrated again and again how important a searching analysis of concepts is for the progress of science. A change of concepts, or a drastic reformulation of existing concepts, may lead to a veritable scientific revolution.

The nature of scientific revolutions, as you are probably aware, is a rather controversial subject. T. S. Kuhn proposed the thesis that one can distinguish in science long periods of even progress, separated from each other by rather violent scientific revolutions, during which one set of concepts is replaced by an entirely new set of concepts or a new paradigm, as it is called by Kuhn. This description of scientific revolutions may be valid for the physical sciences, but it is becoming steadily more apparent that it is not a correct representation of what usually goes on in the biological sciences. Let me illustrate this by one example. When Darwin published his new theory of evolution through natural selection in 1859, it by no means replaced the Lamarckian interpretation of evolution, and not even the rise of genetics in the early 1900s was able to do this. A glance at the prin-

Ernst Mayr · Museum of Comparative Zoology, The Agassiz Museum, Harvard University, Cambridge, Massachusetts 02138.

cipal biology texts in this country, Britain, and the European continent during the 1920s and 1930s proves conclusively that at that time Lamarckism was still very much alive and kicking. What actually happened in evolutionary biology was a synthesis in the 1930s and 1940s of the best, the most viable, elements of a whole series of competing theories and traditions.

I could give you a description of similar theory formation in several other areas of biology: two and sometimes three or even four competing theories fuse in such a way that each of them contributes its most viable parts while abandoning others. Indeed many of the other great controversies in the history of biology were terminated not by the outright victory of one of the competing dogmas but by a fusion of the best components of both. It seems to me, as somewhat of an outsider, that the study of animal behavior is at the present time in the process of undergoing a similar synthesis.

I cannot describe what is involved without going back into the history of this field—and this history is a venerable one, ultimately going back to Aristotle and the Hippocratics.

Among the historians of science one can recognize two opposing views: the externalists and the internalists. According to the externalists, science is part of the total cultural stream, and developments in science, including scientific revolutions, are vitally influenced by social, economic, religious, psychological, and artistic forces. Their opponents, the internalists, believe that scientific ideas have a life of their own, that they are affected primarily by discoveries within the field, and that new concepts are derived directly from a study of the evidence. As happens so often in long-drawn-out arguments, there is some truth on both sides. In some branches of biology unquestionably there is hardly any evidence for external influences. Almost all of our recent advances concerning the understanding of cytoplasm are, for instance, due to the invention of the electron microscope. There are many other branches of biology for which a similar predominance of internal factors is true. Yet the externalists can make a good case for the existence of strong external influences on the development of the evolutionary theory, and, as I may be able to demonstrate later, this is true even more for the study of animal behavior, at least in past centuries.

If the study of animal behavior, or at least theory formation in this field, is now in the process of a synthesis, we should ask the following three questions:

1. What are the various competing theories?
2. What misunderstandings were the cause of past controversies (for when ultimately resolved it is almost invariably shown that heated controversies were based on misunderstandings)?
3. What factors favor the eventual synthesis?

So far as space permits I shall lead up to the answering of these three questions.

Let me begin by laying the foundation through the removal of one of the most fundamental and most damaging misunderstandings—the problem of causation in biology. In the physical sciences we deal invariably with only one set of causes. There may be several causes acting simultaneously, but they have equivalent effects. In biology, in contradistinction, we are always dealing with two sets of causes, proximate and ultimate. What do I mean by this terminology? Let me try to make this clear with the help of an example. What is the cause of bird migration? If a particular bird individual has been with us all summer but goes south in the night from the twenty-sixth to the twenty-seventh of September, we know that some physiological process must have been going on inside this bird. There must have been some interaction between sensory input, hormonal regulation, and response of the nervous systems which induced migratory restlessness in this bird. Further analysis will show that photoperiod (decreasing day length) and perhaps a sudden drop of temperature or change in barometric pressure triggered the actual departure. All these purely physiological aspects of behavior are *proximate* causes. They are the causes responsible for a particular activity in a particular individual at a particular moment. But the little screech owl that lives in the same orchard as the warbler that departed in the night from the twenty-sixth to the twenty-seventh of September is exposed to the same decreasing day length and sudden drop in temperature but does not migrate and, indeed, remains throughout the winter. The reason, of course, is that it has an entirely different genetic program, which through millions of generations has been selected for sedentariness, while the genetic program of the warbler has been adjusted to migration. Genetic programs, their adaptive values, the factors that select for changes in the genetic programs, and the selection pressures exerted on the phenotypes produced by these genetic programs—all these are what we refer to as *ultimate* causes.

When we investigate the research interests of particular biologists, we find that almost invariably they are either students of proximate causes or students of ultimate causes. If they are students of proximate causes their interests are almost entirely confined to the physiological aspects of biological phenomena; if they are students of ultimate causes they concentrate on questions dealing with evolution and adaptation. The students of proximate causes primarily ask questions beginning with the word *how*, while the students of ultimate causes generally begin their questions with *why*. What is often forgotten and must therefore be stated most emphatically is that we do not have a full understanding of any biological phenomenon until both the proximate *and* the ultimate causes have been explored and determined. The physiological explanation must be supplemented in each case by the evolutionary explanation. Or, to state it in terms of DNA, we want to know not only how the genetic program encoded in the DNA is decoded but also what the history and the selective value of a given DNA program is.

If you analyze carefully the arguments between various competing schools in the history of animal behavior, you will find remarkably often that one of the competing camps was interested only in the proximate factors and thought that this represented the entire solution of the problem; the other, in a similar manner, was interested only in the ultimate causes.

After having laid this foundation let me go back to the historical situation. If we wanted to be thorough, we would have to start our history of the study of animal behavior with Aristotle. The two aspects of behavior in which he was particularly interested were the causes of behavior (in a physiological sense) and the differences between man and the animals. Although Aristotle ascribed a rational psyche, true reason (*nous*), only to man, nevertheless the break between animals and man was not nearly so drastic with Aristotle as with Descartes, who represents the real starting point of the history of psychology.

Almost 2000 years had passed between Aristotle's and Descartes's writings, and very few students of animal behavior had lived in the intervening period. None among these is more remarkable than emperor Frederick II, who in his *De Arte Venandi*, a falconry book written around 1245, carefully describes imprinting and shows that the training of falcons is done largely by operant conditioning. His manuscript, unfortunately, remained unknown until rather recently and had no subsequent impact.

An entirely new tradition was started with Descartes. For him there was a drastic difference between man and the animals. Only man had a soul (with the properties of intellect). All other aspects of behavior, such as are found in animals, are functions of the body, and the body as a whole functions like a machine. The properties of this machine "follow naturally entirely from the disposition of the organs, no more nor less than do movements of a clock or other automaton from the arrangement of its counterweights and wheels." Since this machine "comes from the hands of God, it is far better ordered with a far more wonderful movement, than any machine that man can invent." It was this idea, as we shall presently see, which was taken up by the natural theologians and which played such an important role in the history of the instinct concept.

One hundred years later, in 1749, Buffon defended a strictly Cartesian interpretation of animal behavior. Within a few years Buffon's interpretation of animal behavior was heavily attacked by Condillac (1755). For him the newborn animal is a *tabula rasa* and every animal acquires its behavior patterns through a trial-and-error type of learning and internal adjustments. Some of an animal's movements are accompanied by agreeable sensations, others by disagreeable ones. By learning to associate appetites and movements with agreeable sensations the animal, through experience, learns to form certain ideas and connect them with a series of corresponding movements.

I have no time to point out in detail how close Condillac was in his thinking to the school of behaviorism. There is a remarkable continuity of concept, even though I doubt that Watson or any other behaviorist had ever heard of Condillac.

Just as Watson, more than 150 years later, was at once attacked by the ethologists, Condillac was attacked by Reimarus, a natural theologian and the first great representative of the instinct theory.

The interpretation of animal behavior was very different for the natural theologians. They were forever searching for evidence in favor of design, and nothing could demonstrate the existence of a wise and benign creator more conclusively than the marvelous adaptations of animals and plants. Among these the remarkable instincts of animals are particularly convincing, and this is why Reimarus, the leading German deist of his time, chose to write a book on animal instincts. In addition to his arguments in favor of design, Reimarus argued specifically against Condillac's claim that all animal behavior is based on acquired experiences. Not only does Condillac fail entirely to provide any evidence for his claims, says Reimarus, but the known facts are all opposed to them. Except for the relatively few species of mammals and birds with parental care,

> all the other kinds of animals procure their needs from the very beginning without the help of their parents, such as many amphibians, all fishes, and all species of insects, and they are born with the fully developed facility to move in their proper element and to use their extremities in the appropriate manner.

He calls attention to the essential uniformity of species-specific constructions made by animals and refutes the claim that practice is involved:

> One does not find at all in a contemplation of these works that animals in their first tries produce rather imperfect works which become more perfect after much time and practice. Rather they are masterpieces from the very beginning. Some of them are constructed only a single time in their entire life. And yet they would lose their lives if they could not carry out the activity from the beginning in a finished and masterful way. . . . He who studies nature will soon confess that the instincts of animals are not abilities invented by themselves through reasoning and acquired by practice.

We find essentially the same argument in the writings of Paley and Kirby and even in Altum more than 100 years later. The key word in all their writings is the word *instinct*, and, as we know, this word is still with us and was in the title of the first comprehensive treatment of ethology in the English language, Tinbergen's *Study of Instinct*.

Perhaps no concept has played as large a role in the science of animal behavior and has been as controversial as instinct. Even though the great Scholastic, Thomas Aquinas, used the word more than 100 times in his *Summa Theologia*, it was only from the 17th century on that the usage became at all common.

Instinct is one of the many concepts in biology which originated from a "personification," so to speak, of certain activities or manifestations of living processes. The words *life* and *mind* are other examples. Actually, none of these personifications exists as a separate entity, as is implied by the use of the noun. At best there may be instinctive locomotor coordinations, instinctive reactions to certain stimuli, but there is no such thing as *the instinct*. A failure to recognize this led to a wrong formulation of the instinct problem for hundreds of years. To make matters worse, a tradition developed that increasingly used the instinct concept in order to dramatize the unbridgeable gap between animals and man. Man acted appropriately because he had reason, while animals acted appropriately because they had instincts.

I mentioned earlier how strongly theory formation in the study of animal behavior was dominated by external influences. The application of the instinct concept is clear evidence. The theologians and all others who wanted to stress the unbridgeable gap between animals and man denied to animals any intelligence or reason and to man any instincts. Those who wanted to minimize the gap went to the other extreme. They endowed man with a large repertory of instincts and animals with great powers of reason. This led to a highly anthropomorphic interpretation of animal behavior. This approach characterized some of the early Darwinians when they attempted to emphasize how small the step was from animal to man.

Perhaps not surprisingly, those who stressed the difference between man and the animals were the better observers, and their writings are often still highly pertinent, even though we may reject their ideological framework. The sentimental anthropomorphism of an Alfred Brehm, on the other hand, is quite repellent.

By the middle of the 19th century the study of animal behavior did not appear much further advanced than 200 years earlier, at the time of Descartes. Three major interpretations of animal behavior coexisted. In a considerably simplified manner they might be characterized as follows: First there was the Cartesian theory, according to which animals are machines and behavior consists only of automatic reactions to stimuli. In a remarkably Cartesian manner this theory was revived later in the century by Jacques Loeb in his theory of tropisms. Second, there was the *tabula rasa* theory, according to which all behavior is the result of practice and experience. Condillac had been the most articulate proponent of this theory, but it had other adherents. Watson's behaviorism was clearly in this tradition. Third, there was the instinct theory, according to which each kind of animal has its species-specific instincts that permit it to respond appropriately to all the situations it encounters in life.

The one area where some real progress was being made was in the study of proximate causation. An excellent beginning was made in the nineteenth century through the establishment of neurophysiology and sensory physiology by people like Helmholtz, Fechner, Weber, etc. However, this research in neurophysiology was a long way from making connection with the actual interpretation of the behavior of whole animals.

It was in 1859 that Darwin published his *Origin of Species,* which made a giant step forward in the interpretation of the ultimate causes of behavior. To be sure, Lamarck before him had stressed the importance of behavior for evolution, but his speculations on the con-version of habits into species-specific behavior, besides being wrong, had no heuristic value whatsoever.

Darwin's impact on the study and interpretation of animal be-havior was manifold. First of all he submitted massive evidence that behavioral characters evolve exactly like structural characters. This amounted to a refutation of the thesis of the natural theologians that instinct is both so complex and so perfect that it could have come into existence only by a single act of creation. What Darwin showed in the seventh chapter of the *Origin*, which is entirely devoted to a dis-cussion of instinct, is that instincts are variable, that often there are only very minor differences between the instincts of closely related species, and that different modifications of instincts have different adaptive values in different environments. If instinctive components of behavior are as variable as claimed by Darwin, they are able to respond to natural selection exactly as structural characters do. Hence natural selection should be able to influence the evolutionary trend of instincts. Darwin refused to give a definition of instinct, but he said:

> An action which we ourselves should require experience to enable us to perform, when performed by an animal, more especially by a young one, without any experience, and when performed by many individuals in the same way, without their knowing for what purpose it is per-formed, is usually said to be instinctive.

Behavior is treated by Darwin again and again; it occupies a major portion of the *Descent of Man* (1871) and is the whole subject matter of *The Expressions of the Emotions in Man and Animals* (1872). As fascinating as it would be to say a great deal more about Darwin's contribution to our understanding of animal behavior, I shall limit myself to a single point.

The prevailing interest among the students of behavior in the Cartesian tradition was in proximate causes. Their principal question was: "How does this or that behavior function?" Such questions as, "Why does a particular animal or species of animal have a particular behavior and not another one?" were never asked. The natural theologians, in contrast, asked this question but answered it in the most simple-minded way: Animals behave in this manner because it has thus been ordered by God. One of the significant contributions made by Darwin is that he also asked the "Why" question but answered it in a strictly scientific manner without recourse to supernatural causation. An animal acts in this manner and not in another one because such behavior is "adaptive," that is, because the evolution of such behavior had been favored by natural selection.

I shall not bore you with the hundred years of arguments that followed Darwin, during which the philosophers argued whether or not it is legitimate and scientific to ask "Why" questions—nor with the problem of whether or not Darwin's argument is teleological and thus inappropriate and unscientific. It is only in the last ten years or so that a younger generation of philosophers of science have shown conclusively that there is nothing objectionable or unscientific in adaptational language and in teleonomic explanations. I hope to be able to come back to this problem.

Curiously, the new research program that Darwin made possible by his adaptational explanations was used only very slightly by naturalists and students of behavior. Julian Huxley, in his great crested grebe study, was one of the outstanding exceptions until the rise of ethology. Even in ethology, with its enormous interest in evolution, the original emphasis of Whitman, Heinroth, and Konrad Lorenz was on reconstructing phylogeny rather than on demonstrating the selective value and adaptive nature of genetic behavior programs. But this has changed drastically in recent years, particularly at the urging of Niko Tinbergen.

Let me now take a different tack. Darwin had not yet entirely freed himself from a belief in an occasional inheritance of acquired characters. He thought that habits could sometimes become instincts in the course of many generations. This alternative to strictly inherited behavior was very popular among the social reformers, who were so fervent about improving man himself by improving man's environment. This was one of the many reasons why Darwinism was at first so popular among Marxists, particularly Marx himself and Engels. The situation changed drastically, however, in 1883 when Weismann emphatically rejected any possibility of an inheritance of acquired characters and hence any possibility of a conversion of habits into instincts. Darwinism lost much of its attraction for the Marxists as a result of Weismann's thesis, and part of the reason for the temporary popularity of Lysenko among certain groups of psychologists was that they thought this would restore the pre-Weismannian thesis of the convertibility of habit into genetic properties. On the whole, however, Weismann's thesis spread rapidly among the students of animal behavior, particularly through Lloyd Morgan's efforts, and after the 1890s hardly any serious author maintained a belief in an inheritance of acquired characters.

The result was an enormous popularity of instinct in the first decades of this century. Alas, to a large extent it was a rather uncritical acceptance of the concept of instinct and its application to entirely unsuitable phenomena. Worst of all it had a stultifying effect because it favored the erroneous assumption that one had provided an explanation if one had attributed a phenomenon to instinct.

There were two reactions against this sweeping and uncritical application of instinct. The behaviorists, and indeed the majority of authors limiting themselves to a study of proximate factors, simply eliminated instinct from the repertory of their explanatory concepts. Either they did not realize it or else they did not care that this eliminated all explanation of the ultimate causation of behavior. The ethologists, in spite of much fumbling and steps in wrong directions, adopted a more constructive attitude by undertaking a dissection of instinct. What do I mean by this? In the first half of the 19th century the great polarization was still *instinct*, which was a property of animals, versus *mind*, which was a property of man. All endeavors to study "the instinct" were just as futile as those to study "the mind." Even though Tinbergen, as late as 1950, entitled his book *The Study of Instinct*, he actually does not deal with the instinct anywhere in this book. What he deals with are instinctive responses to signals and instinctive displays. The dissection of instinct by the ethologists led to a tripartite structuring of instinctive behavior: the cognitive part, such as the response to signals or releasers; the conative part, the readiness or refractoriness of responding to such signals; and the executive or locomotory component of the behavior, that is, the display or other response to the stimuli.

In each of the three components of a total behavior pattern—and of course the more recent studies undertake a much finer subdivision than one of only three stages—there may be what used to be called instinctive components. With the word *instinct* having become so ambiguous, the word *instinctive* was more and more replaced by the word *innate*, or genetically based.

This by no means ended the argument and intellectual confusion. Indeed, as you will all remember, there have been few controversies in the history of the study of animal behavior more heated than the one dealing with the question of what is innate and what is acquired.

At the risk of greatly oversimplifying the issue, one might state that the students of proximate causations tended to explain behavior largely as being acquired, while the students of ultimate causation tended to interpret most behaviors as innate. Or, as Lorenz later expressed it, "as that which we formerly called innate."

I do not think it has ever been pointed out that the polarity of innate versus acquired is scientifically meaningless. *Innate* refers to the genotype, while *acquired* refers to the phenotype. It is as if someone said that a given character of an organism was part of its phenotype and somebody else said it was part of its genotype.

To get away from this ambiguity and to make use of the enrichment of our scientific vocabulary provided by information theory, I proposed in 1965 to make a distinction between closed and open programs.

Let me explain this with the help of an example. When a cowbird female lays her egg into the nest of a song sparrow and the song sparrow parents raise the young cowbird, this young bird has never seen another cowbird. Yet two or three weeks after it has left the nest, having been fed by his song sparrow parents during the entire time, the young cowbird suddenly leaves the song sparrows and goes in search of other cowbirds. It joins one of their flocks, stays with them throughout the fall and winter, and finally, in spring, mates with another cowbird. Here we have what I call an entirely closed genetic program. As far as mate selection is concerned, at least as far as the nature of the species of the mate is concerned, there is no input through experience. A similarly closed program dominates the life histories of most short-lived species, which includes more than 95% of all organisms. A female moth may mate within twenty minutes after she emerges from her chrysalis, and if she makes a mistake in selecting her mate she will leave no offspring, a fact which was used so brilliantly in the eradication of the screw worm.

Let us now take an entirely different example. In species in which there is imprinting, so charmingly described by Oskar Heinroth or Konrad Lorenz, the young hatchling bird has all the innate behavior pattern of following a parent, but the diagnostic characters of the parent that is to be followed are supplied during the first few hours after hatching. Such imprinting works only when there is what I call an open genetic program. How firmly this information can be imbedded in the program was demonstrated by Immelmann for estrildid finches, in which the pair bond can no longer be broken once a male has been imprinted on a different species.

Let me frankly admit that we are not yet entirely out of the woods. When I proposed the terminological improvement of closed and open genetic programs I was less than precise. I had again forgotten about the difference between genotype and phenotype. What controls a particular act of behavior is not the genetic program, as such, but a behavior program in the nervous system which in turn is the translated product of the genetic program plus whatever inputs have been contributed by experience. When dealing with the interpretation of behavior we should, therefore, not refer to open and closed *genetic* programs but rather to open and closed *behavior* programs. It is, of course, obvious that the behavior program which is translated from the genetic program has many open areas into which information can be inserted as the result of experience.

Where we go from here is a matter of personal interest. The physiologist, in other words the student of proximate causation of behavior, will attempt to determine how behavior can be translated from the genetic program into the behavior program of the nervous system and how additional information can be incorporated into the behavior program of the nervous system. This includes a study of all such problems as the ontogeny of behavior, the role of practice in behavior, reversibility of behavior, imprinting, temporary learning, and a host of similar topics. The role of endocrine factors in the evoking of particular behavior manifestations is a major area in these studies. And this of course was Danny Lehrman's major field of interest. Frankly speaking, I am not qualified to speak on this subject. However, fortunately, Danny was also an ardent evolutionist and very much interested in the adaptive aspects of behavior, and for this reason I might be permitted to add a few further comments.

What are the behavioral problems that the evolutionist is interested in and what kind of questions does he ask? There is first the old Darwinian question of whether different kinds of behavior really have different selective values and can be shaped by natural selection. Niko Tinbergen and his school have done a good deal of work in this area and have demonstrated quite conclusively that even rather slight differences in behavior may often be of considerable selective significance.

Second, let me now take up an entirely different evolutionary consideration. Knowing that some behaviors are largely controlled by closed behavior programs, others by open ones, we can ask for what kinds of behavior and under what particular circumstances an open behavior program would have a greater selective value than a closed one? In an attempt to find answers to this question I classify behavior according to a number of purely pragmatic categories.

Communicative behavior among individuals of the same species involves the production of signals by one partner and the reception of a piece of information, a message, by the other. The interactions between male and female, between parent and offspring, and between rival males are all more or less in this category. What do we find about the nature of the behavior programs?

The courtship between males and females in animals has, among other functions, that of preventing hybridization. In other words, the displays serve as behavioral isolating mechanisms. Since mating in most short-lived animals occurs only once in their life-span and often within hours, if not within minutes, after hatching, the behavior program must be rigidly closed. I do not know how often it has been tried, but to the best of my knowledge no one has ever succeeded in conditioning the male of a species to perform the courtship display of a different species. There is a possibility for this in birdsong, but the functional interpretation is ambiguous.

A second set of intraspecific communicative behaviors relates to parent–offspring behavior. Among the relatively small percentage of animals with parental care we find a wide range of behavior programs from almost completely closed ones to almost completely open ones. Pheasant and turkey hens kill chicks of species other than their own, governed by a closed program, while females of other species readily accept nonconspecific young. Imprinting plays a large role in this area and is sometimes largely irreversible; that is, the open genetic program is converted by the imprinting experience into a closed behavior program.

What is particularly interesting is that in many species, particularly in social species, there is, not surprisingly, a selective premium on individual recognition. Among seabirds that live in large colonies and also among herd-forming ungulates, both mother and offspring learn to recognize each other individually. To the observer it almost seems like a miracle that parent and offspring should be able to recognize each other instantaneously in a seabird colony consisting of 10,000 or 100,000 individuals, but they do.

As a broad generalization, one can state that the shorter the life cycle, the less reliance there is on open programs. On the other hand, the longer the period of parental care, the more information can be stored in an open behavior program. This evolutionary trend of course culminates in man.

Closed programs play an important role not only in the interaction of individuals belonging to the same species but also in interspecies interactions. Of the rich variety of phenomena in this area let me mention only three categories: (1) mixed-species aggregations, such as mixed birdflocks or herds of ungulates; (2) various types of symbioses; and (3) all kinds of predator–prey relationships.

In all these interactions the exchange of signals plays an important role. These signals, at least in species without parental care, cannot be learned by the young individuals; they must have a ready answer available for the most important encounters with other sympatric species as soon as they are born. Closed genetic programs, therefore, play a considerable role in interspecies interactions.

The situation is, however, quite different for *noncommunicative behavior*. I am referring particularly to behavior that plays a role in food selection and habitat selection. In contrast to communicatory behavior, such behavior is largely controlled by open programs. There is a considerable selective premium on flexibility, and great scope for learning and imprinting. Salmon fry, for instance, are imprinted on the particular chemical properties of their native stream and will try to return to this stream for spawning. It is very often possible to condition a young animal to a new kind of food or to a different habitat. But closed genetic programs may play a large role even in noncommunicative behavior. Many host-specific insects cannot be conditioned to a different host. Virtually all construction behavior, whether it is the weaving of a spider's web or of the chrysalis of a moth, or nest building by a weaver bird or by a wasp or a bee, is strictly controlled by a closed genetic program. A strict coordination of movements is indeed of high selective value, considering the great survival importance of the structures that result from the construction behavior.

Let me add just a few words about a third kind of interest of the evolutionary biologist. Behavior not only evolves, but is itself one of the most important selection pressures we know. Virtually all major evolutionary developments—whether the coming on land of previously aquatic evolutionary lines or the utilization of an entirely new habitat or an entirely new source of food—are the results of behavioral shifts. There are reasons to believe that such important behavioral shifts often take place in open behavior programs. Naturalists have often described the acquisition of new behaviors by certain individuals of a species. The famous opening of milk bottles by British titmice is a well-known example. If such a nongenetic shift of behavior permits the exploitation of a previously unused resource of the environment, selection will tend to reinforce this deviant behavior genetically.

In contrast to this, communicative behavior seems to play only a small role in evolutionary shifts. This is not surprising. Noncommunicative behavior is directed toward the exploitation of natural resources and it should be flexible, permitting an opportunistic adjustment to rapid changes in the environment; and it should also permit an enlargement of the niche as well as shift into a new niche. This is possible only if such behavior is largely controlled by an open program.

The great difference between communicative and noncommunicative behavior explains some of the controversies in the recent behavior literature. The ethologists have been primarily interested in species-specific signals and their evolution. Comparison of different species has been of great concern to them. The classical experimental psychologists, on the other hand, almost invariably work only with a single species, and their primary interest has been in learning, conditioning, and other modifications of behavior. The phenomena they studied were to a large extent aspects of noncommunicative behavior, like maze running or food selection. The two groups of investigators dealt with different kinds of behavior and there can be no doubt that behaviors can be very different when classified according to the target of the behavior and to the particular selection pressure to which it is exposed.

Behavior constantly interacts with both the living and the inanimate environment and is thus constantly the target of natural selection. In order to provide the optimal response to these pressures, it is sometimes advantageous for the genetic program governing the behavior to be largely closed, while in other behavior interactions and in other types of organisms an open behavior program is favored by selection.

The study of animal behavior, like almost all branches of empirical science, has suffered in the past from too ready an acceptance of certain terms like *instinct* or *learning* as complete explanatory designations of highly heterogeneous sets of phenomena. The field has likewise suffered from an insufficient analysis of the underlying concepts. Too many students of behavior have acted like the proverbial blind men and the elephant. They got hold of one particular aspect of behavior and thought that it would permit the explanation of all behavior. We now know that it will not. What we need is a broad-gauge approach, an approach that is tolerant and open-minded toward the explanatory models of other workers in the field who happen to have gotten hold of a different part of the elephant. Only that will eventually lead to an understanding of the elephant in its entirety.

It would seem particularly appropriate to make this point in a book dedicated to the memory of Danny Lehrman, for he indeed had an open mind, appreciated most of the different currents in the study of animal behavior, and was earnestly concerned with producing a true synthesis. Let us continue our work in a spirit of tolerance with willingness to revise our concepts whenever new facts or insights show them to be invalid.

2

On the Integration of Gender Strategies in Mammalian Social Systems

J. H. Crook

At a time when the nature of the relationship between men and women and of the role of women in modern society is undergoing continuous reevaluation, it is perhaps important to see these issues in whatever light comparative biology can throw on the problem. Without for a moment denying the uniqueness of man and the complexity of our patterns of interpersonal sexuality, we know that our species has undergone a biological evolution in many respects quite as important to our social origins as the vicissitudes of cultural history. A number of recent papers have tended to stress this theme and to balance the heavier weight of studies attributing gender contrasts purely to cultural acquisition on learning. In that this tendency in the literature broadens our perspective as a whole, it is, I believe, much to be welcomed. One way the comparative biologist can assist is by attempting to describe the range of gender relations found in the order of mammals, to which we belong. Such an attempt leads naturally to an endeavor to explain the differing types in terms of their functions within the lifestyle and ecology of the species concerned.

When my thoughts began to turn in this direction, I was immediately struck by the fact that, in by far the majority of extant mam-

J. H. Crook · Department of Psychology, University of Bristol, United Kingdom.

mals, relations between the sexes are trivial in the extreme. Often males and females seem to exist almost as separate species with spacing and activity patterns quite distinct from one another, each with markedly different biological functions with respect to reproduction, and a brief act of mating being the only point of contact in otherwise separate lives. It is as if two quite contrasting life cycles were only geared together at one small, tenuous, but critical point.

Personalized sex, by which I mean long-term relating between individual animals in the mutual task of consortship while mating and in rearing their young, is rare and clearly arises only under particular conditions. Since our own case belongs to this category, we may reasonably wish to know what biological conditions appear responsible for its emergence.

Reproductive Strategies and the Mating–Rearing System

In the older literature on mating–rearing systems (M–R systems) there often appeared to be a tacit assumption that males and females share a common aim in maximizing the effective reproduction of their generation and that their association comprised an optimal strategy to that end. In a certain vague sense this notion is true enough, but it neglects certain essential facts regarding the relative involvement of the two sexes in reproductive activities. Among mammals, the female, by virtue of pregnancy and lactation, spends much more time and energy than the male in bearing and rearing the young. Indeed this is but an extension of the primary differentiation between the sexes in the animal kingdom as a whole, in which the biological investment in eggs is normally greater than that involved in the production of sperm. The contrasts between the relative investments of each gender in producing and rearing a young offspring imply that the substrategies of each gender are not the same, even though both subserve the overall strategy of optimizing reproductive success. The interaction of the two sexes that comprises any given mating and rearing system (M–R system) is thus an expression of *gender-specific* strategies for ensuring gene transmission into future generations.

In common with that of certain other vertebrates, the reproductive strategy of mammals may be made up of three successive phases:

1. The precopulatory or courtship phase.
2. The postcopulatory relationship phase.
3. The parental or rearing phase after the young are born.

Behaviorally the three phases are mediated by rather differing motivational or affectional systems. In the courtship phase two animals approach and adjust motivational responses to one another in such a way that copulatory behavior is disinhibited and facilitated to full expression. In the second phase bond-maintaining behaviors mediate the continuing association of the pair, while in the rearing phase the activities of the animals are more exclusively directed to the care of young. Females are necessarily involved in all three phases or, in cases where the male does not remain with her after mating, in at least the first and last. Males, however, vary enormously in the extent to which they appear. They are necessarily present in the first but not necessarily in the other two. In fact, as we shall shortly be able to see, approximately 60% of M–R systems in a recent mammalian listing involve the male no further than in copulation. In a further 10%, the male's presence does not contribute specifically to the rearing of his own young, and only in some 30% are males bonded over long periods of time with particular females; and in these he participates in various ways in caring for the young in perhaps only about 15% of systems. In general then, female mammals are predominantly involved in rearing and caring while males are predominantly concerned with copulation. Paternal care and long-term bonding is uncommon.

These matters have been examined in depth recently in a valuable article by Robert Trivers (1972). He examines the involvements of the two sexes in terms of their respective time and energy investments in their offspring. Trivers defines *parental investment* as "any investment by the parent in an individual offspring that increases the offspring's chances of surviving and hence reproducing at the cost of the parents ability to invest in other offspring." In Trivers's terms, therefore, a large parental investment is necessarily one that decreases the parents' ability to produce further offspring. From this starting point a number of somewhat axiomatic statements follow.

First, natural selection may be expected to favor the sum of parental investments in offspring per breeding season that leads to maximum net reproductive success. As a result of pairing, the total number of offspring arising from one parent equals that from the other. Yet the male, by mating with more than one female, may contribute to more offspring than those arising from one particular pairing. This means that the male's reproductive capacity is limited by the female's reproductive investment and that females thus become the resource limiting the productivity of males. Further, it follows that individuals of the sex that invests less in parental activity will tend to compete with one another for those who invest more, since, by successively pairing with the sex giving the greater investment, the individual can increase his overall contribution to the following generation.

Since, among mammals, it is characteristically the male that invests least parentally, we may expect to find that males are typically very much concerned with mating with several females and with excluding other males from so doing. This competition is of course the process described by Darwin as sexual selection and, in particular, intrasexual selection among males (see, for example, B. Campbell, 1972). As Trivers (1972) points out, the process is very largely governed by the relative parental investment of the sexes in their offspring. The outcome of sexual selection is characteristically the emergence of psychoendocrine mechanisms favoring male success in male-to-male competition supported by morphological characteristics of display and weaponry functioning to the same end. In addition, sexual selection commonly produces physical dimorphism and larger-sized males than females. In behavioral terms we may expect to find male mammals adopting strategies of rapid successive mating and other-male exclusion. Female strategies will be far more concerned with ensuring copulation with males of what we may loosely call strong genetic constitution—that is to say, those whose behavior and appearance are indicators of their siring highly viable offspring. Besides following those strategies, females may be expected to concern themselves primarily with caution in self-preservation when they are carrying embryos in pregnancy and with nurturing and care-giving behavior as the young grow after birth.

Situations arise, however, in which the male's tendency to maximize matings may be curtailed. For example, a male may have a greater chance of siring young if he mates only with females clearly in estrus and spends his time near such animals while excluding other males from them than he would by disseminating his sex products among any or all of the females he encounters. He may also increase his chances of siring if he coerces a group of females to remain with him for a time period sufficiently long for all of them to have come into estrus. However, should he remain with some or all of these females beyond that time, his chances of further sirings will be reduced. He is likely therefore to try to maintain some throughput of females so that his rate of successful impregnations remains as high as possible. These considerations apply purely to the question of maximizing pregnancies and thus primarily concern the first phase of the overall mammalian scheme.

I have pointed out elsewhere (1972) that one male's success relative to others ultimately depends on the survival of his offspring to maturity. It thus follows that should a male's participation in rearing or caring activities increase the probability that his young will grow to reproduce themselves then it is likely that selection will favor changes increasing such participation. The balance between the effects of mating with more females or, alternatively, mating with fewer but helping some more, may often be quite delicate, and doubtless this has always been so during the progressive evolution of mating–rearing systems. As Trivers (1972) also points out, male parental investment will be particularly favored by selection when his increased interest in mothers and young is more reproductively effective in the long term than mating with a longer string of unassisted females. Here then is the prime condition under which relating between individuals of the two sexes is likely to have emerged in the several mammal groups in which we find it (see further Goss-Custard, Dunbar, & Aldrich-Blake, 1972; Orians, 1969).

Sociotypical Radiation and the Adaptive Significance of Mammalian Mating–Rearing Systems

In a recent comparative survey of dispersion and grouping types of mammals, some colleagues and I attempted to categorize the different types of mating–rearing systems into which any mammalian species could be placed (Crook, Ellis, & Goss-Custard, 1976). Mating–rearing systems were one aspect of the range of sociotypes considered and the only one we need treat in some detail here.

The categorization was derived from an extensive reading of available literature, starting from J. Eisenberg's important survey of 1965 and working through more recent studies. Each order was separately examined, and the number of distinguishable sociotypes listed. When the lists for each order were compared it was apparent that between those orders that showed a considerable sociotypical diversity there was a marked similarity in the categories employed. If one considers parallelisms as being adaptations to similar life conditions, then one is led to suppose that repeated convergence in socioreproductive structuring has occurred during the adaptive radiation of the class as a whole. A total of some 16 sociotypes covers the range of organizations shown in the class.

In Table I the number of sociotypes listed per order is shown and their distribution among five types of mating–rearing systems il-

Table I. Distribution of Sociotypes of Each Mammalian Order across Mating–Rearing Systems

		Number of sociotypes				
		A Brief copulatory	B	C	D	E Diffuse multimale
Order		association	Short-term polygyny	Long-term polygyny	Monog- amy	grouping
Total available sociotypes:	16	3	5	2	2	3
Marsupials	6	1	1	2	2	—
Insectivora	5	3	1	1	—	—
Chiroptera	2	—	—	1	—	1
Primates	6	1	1	2	1	1
Lagomorpha	3	1	2	—	—	—
Edentata	1	1	—	—	—	—
Pholidota	1	1	—	—	—	—
Rodentia	7	3	2	1	1	—
Carnivora	5	1	2	—	2	—
Pinnepedia	2	1	1	—	—	—
Perrisodactyla and primitive ungulates	3	1	1	—	—	—
Artiodactyla	7	1	2	2	—	2
Sum of sociotypes:	48	15	13	9	6	5
Percentage:	100	31.3	27.1	18.7	12.5	10.5

lustrated. The five mating–rearing systems are as follows:

(A) Well-spaced solitarily living individuals or mother–litter units with male home ranges usually larger than those of females and including several. Males with ranges covering those of females tend to exclude subordinate males to peripheral positions. The males are commonly very mobile in their home ranges and visit their females regularly but do not associate otherwise with them. They give no assistance in rearing. Examples are many insectivores such as shrews, primates such as *Microcebus*, and gerbils (*Meriones*) among rodents.

(B) Males associate with several females for mating purposes either synchronously or in overlapping succession. The apparent "harems" are not thus of constant membership but contain a steady flowthrough of females attracted by, but not otherwise involved with, the male. Such systems occur statically in colonies based on refuges, in herds where males establish territories in areas grazed by females, or in nomadic congregations (e.g., seals, deer).

(C) Long-term polygamous units develop when true bonding maintains links between males and members of a true "harem" over long time periods. Both this system and the one above necessitate a population of nonbreeding adult males variously dispersed in relation to the reproductive units (e.g., horses, geladas).

(D) Monogamy occurs in a few species. Male and female live together in a territory, and both participate in rearing young. In carnivores parents may associate in small hunting clans (e.g., coyotes, marmosets).

(E) In large herds males may relate to one another in some form of a dominance hierarchy only the more senior members of which have access to females in estrus. Consorting is relatively short and males tend to mate with several females. There is no paternal care in the strict sense since paternity is uncertain, but collectively males show considerable caring behavior and protective activities in relation to both females and young. Dominant animals tend to regulate affect in the group by breaking up quarrels, etc. (e.g., *Papio anubis,* chimpanzee).

A glance at Table I shows that Mating–Rearing System A is the commonest across the widest range of mammalian orders. It is particularly well developed among small-bodied mammals—insectivores and rodents. Its widespread occurrence among mammalian stem forms as well as its wide representation in derivative orders suggests that we have here something close to the primordial mammalian mating system. Here the separation of the sexes is maximal, with the male's sole interaction with females apparently being copulation and regular checkups on their sexual condition. Males are otherwise occupied with various types of exclusive activities whereby members of their own sex are denied access to the females they cover in their ranging.

The widespread trend toward short-time polygyny is indicated by its occurrence in all but two orders. This suggests that polygyny has repeatedly evolved in a variety of contrasting forms under the influence of similar conditions. While it would need a book to detail the range of various organizations represented by the different orders, and while it is certainly dangerous to attempt generalizations that can be expected to account for them all, certain principles seem to hold here. Short-term polygynous sexual associations occur in animals as diverse as wildebeeste, bats, and pinnipeds. The association is short in that, as soon as mating is completed, the male is likely to turn his attention to other females, whether his most recent mate remains with him or not. Many ungulates of the African plains are dispersed in such a way that female herds occupy the most food-rich areas. Here too the males congregate and, when in reproductive condition, they endeavor to round up females into a relatively coherent group. The extent to which they succeed varies greatly across species and the time for which the association is held is also a variant. The male's strategy is, however, clear; he is basically holding a stock of females for mating purposes and allows a viscous flow of members through his unit. The strategy is thus one of maximizing copulations through establishing a holding unit for the purpose. This is the most economical way of operating where animals are in relatively dense congregations. The regulation of behavior imposed by the system probably conserves energy that would be wasted in random and frequent contests and chases that would otherwise ensue in the herd at large were a maximizing strategy without the development of a holding unit to be attempted.

Such a system as Mating–Rearing System B seems commonest

among animals that congregate. Its difference from Type A may thus be expressed primarily in spatial terms. In A, males attempt to regulate other-male access to females living solitarily in relatively fixed ranges, and they do this by covering such ranges with their own mutually exclusive larger one. Where individuals associate in colonies, males attempt the same strategy, but this time the consequence is a herding of females into groups that have the appearance of harems but which have a changing memberhsip. Estes's (1966) work on the wildebeeste shows that the question of whether this harem is localized within a topographically defined home range or not is a function of local feeding ecology. In general, when feeding conditions are good and remain so for long periods group territories develop with males particularly concerned with holding ground against other males and with gaining females. Where conditions are unstable the male is mobile, performing herding activities within a nomadic concourse. Females cannot be coerced beyond a certain degree and, doubtless, exercise choice in pairing dependent on responses to the male's courtship behavior. In an extreme development of this system, the Uganda kob (Buechner, 1961) operates a lek to which he attracts a succession of females for mating. His association with his mates is thus as brief as that of the ruff or an equivalent lekking bird.

Long-term polygyny (System C) occurs where the male and female remain together for periods of time as a result of a bonding process that is mutual. While the extent of the contribution to the bond by each sex differs between species as does its duration, the prime difference from B is that the male in the reproductive state is not necessarily always engaged in herding females and excluding rivals. This system has emerged in some marsupials, rodents, bats, Artiodactyla, and especially in primates (Crook, 1970, 1972). In strategic terms the male is here sacrificing chances of mating unbonded females by remaining with his own group. Probably there are several sets of circumstances, as yet only poorly outlined, that may encourage him to do this. One is a continuation of the trend already apparent. By establishing a bond the male increases the certainty of a pregnancy. Perhaps he is especially likely to take this course if the population commonly fragments into small, separately foraging parties in response to conditions of food-item dispersion. Separately traveling small groups of females gain a large-bodied male protector under such circumstances and this may favor a bonded arrangement. Such males are also likely to preserve the females from a harassment by other males which may be detrimental to rearing young.

Two other conditions may be important: One is where defense of a shared home range reserves food or other resources, maybe a refuge, for the exclusive use of the group. The male's strategy here would be to preserve these resources for those females carrying his own young rather than those of others. Another condition is where the infant is dependent for a long time on the mother, and male assistance in some protective or caring activity increases the likelihood of his own young being successfully reared. These three conditions may of course arise together, so that the balance of selection pressures involved may be complex indeed.

Males showing long-term polygyny are as likely to exhibit marked morphological and behavioral secondary sexual traits as a consequence of sexual selection as are those with short-term polygyny. They are as much involved in other-male exclusion and indeed have to maintain this over long periods. They also have to prevent the takeover of their females by rivals. Their large size, evolved primarily in relation to sexual selection, is also useful in antipredator activities. Females tend to gain protection from harassment by conspecific males, from predators, and from food shortage (in territorial species) as a result of the male's excluding behavior. Insofar as they can determine how well a male may perform in these respects, females are likely to choose good performers.

Monogamous long-term bonding (System D) is relatively rare. Koalas, gibbons, marmosets, coyotes, and similar social carnivores more or less make up the total. The conditions under which this system emerges often seem to involve a long period of infant dependence on the mother and the involvement of the male in food collection and care and carrying of the young. The male here opts for ensuring effective rearing, which owing to the long-term dependency of the child, may be at some risk without his assistance, and he probably achieves this mainly by reducing his mate's energy expenditure. He is less clearly always the protector, which suggests that his role in this respect has arisen through the sexual selection of his practices excluding conspecifics rather than in relation to predation. Monogamous males tend to differ less from females in size and behavior characteristics than do polygynous species. In the marmosets (and certain other South American monkeys) the male is smaller than the female and subordinate to her. Here the male's carrying of the twin young probably allows the female to hunt and otherwise forage less expensively than would be the case if she were having to carry them. She may also be pregnant again at the same time (J. Ingrams, personal communication).

The evolution of multimale reproductive groupings (System E), where brief sexual consorting occurs together with varying degrees of hierarchy formation regulating reproductive access to estrous females, seems to occur where there are strong pressures favoring well-integrated social groups with limited degrees of splitting. Such conditions arise under conditions of high predation and adequate food resources. Here males commonly show activities that are care-giving to a wide array of fellow group members. The advantages of a male's behavior are thus not restricted to his own young but include those of others. Indeed in this type of structure, paternity is not indicated in the social arrangements and only the matrilinial kin can be easily traced. It is usual, however, for the most dominant males to father the majority of young. It is thus in accordance with evolutionary expectations that these same males would show the most group-regulatory and protective behavior. In the chimpanzee the tendency for the group to fragment into food-searching parties seems related to the extreme sexual permissiveness and the relative absence of autocratic hierarchy formation. It seems likely that the extensive regulation of affect in chimp groups is related to these relatively unstructured sociosexual conditions.

Evolution of Parental Care with Particular Reference to the Male

Among mammals litter size probably reflects the number of young that can be raised successfully (Lack, 1954). The same is probably true of the inherited growth rates of young animals and the duration of the period of their dependency on the mother. In the majority of species the young grow and are weaned relatively quickly and are then often forcefully encouraged to disperse beyond the maternal home range. As with birds, this phase of dispersal is probably accompanied by a high mortality, and only a small proportion of weaned young find suitable home ranges for themselves and encounter mates there. In many mammals, males are particularly prone to peripheralization, being pushed out into suboptimal habitats so that most of the reproduction is performed by relatively few males—the others being a reproductively redundant or surplus population. It is a thought worth pondering that even in our own species, wars have from time to time occasioned imbalance in sex ratios among adults. Certainly we do not need all those males around simply to maintain population. Biologically, human males are commonly expendable.

Slow growth and long-term dependency are doubtless related to special needs of the young in relation to acquiring independence without too high a risk of mortality. Clearly in an ideal world females would maximize reproduction by getting rid of their young and rearing new embryos as soon as possible. Delays in this process imply the necessity of care. Probably care is often related to the acquisition of difficult feeding habits, such as those of many carnivores. Here the males may assist the females in catching food for the young while they are learning. The relationship between individuals then produces a complex social environment necessitating social skills as a passport for membership. Young may take time to learn such behaviors. Where the physical environment is complex (for example, branches and twigs in trees), some species may need to learn how to use it, and this may take time. In addition, where young are dependent for long periods, the risk that predators will take them is increased, necessitating means of protection, which encourages social grouping and responses that in turn require a learning of social skills that likewise takes time. It seems therefore plausible that long-term care and social complexity will tend to correlate in comparative studies.

If one compares ungulate, carnivore, and primate systems, perhaps the main point to note is that in ungulates, where for the most part the young grow to independent movement and foraging quickly, short-term polygyny with little male care is common. In carnivores the young need to be raised carefully and fed and taught to feed themselves. Here social systems get very complex, reaching a peak in the elaborate food-sharing rituals of the hunting dog. Among primates, too, long-term maternal dependence goes with male care, although only in some forms does this extend to feeding. Primates seem to specialize in complex social organization, although not to a degree more elaborate than that of some carnivores (Schaller, 1972; Kleiman & Eisenberg, 1973).

In carnivore and primate social organizations, male attention and care of young does not in all cases fit Trivers's definition of parental investment. In particular there are various behaviors which aid young but which do not interfere with the male's investment in other siblings. There are, furthermore, types of behavior in which male care is collective and benefits young animals other than his own progeny. Trivers's definition thus delineates but one aspect of male attention to young—exclusive care of an individual that precludes investment in another young at the same time. This definition, while important for the development of the Trivers model, does not comprise the whole set of care-giving behavior which may yet form part of a male's reproductive strategy. There is, for example, care in which a male protects either individually or collectively the young of his reproductive group. There is also care in which protection is given to some other youngster not clearly related to the male in question or to several at the same time. We can thus distinguish between offspring care and collective caring. Sometimes both types of protection coincide, as when Gartlan (personal communication) observed the males of several patas monkey groups, normally foraging in widely separated home ranges, congregated near a well. As jackals approached, the males coordinated their behavior to drive them away, irrespective of whose reproductive group was nearest to the potential predator. Male primates, both adult and subadult, have commonly been seen to rescue infants, irrespective of whether they live in one-male or in multimale reproductive groups.

Male care (Deag & Crook, 1971) involving sitting with, grooming, and paying close attention to juveniles or babies other than progeny is also common. The extent of such care in primates varies enormously both among species and among populations of the same species. Sometimes, in fact, infants spend more time with males than with any other age–sex category apart from their mothers. In the rhesus, which is not well known for paternal care, males may be attracted to newborns or to distressed infants. In laboratory studies Mitchell, Redican, and Comber (1974) have shown that adult male rhesus monkeys have considerable unused capacities for providing close contact comfort and vigorous sibling-type play with infants, especially male ones. The attachment between the animals was often extremely strong and lacked the rejecting components present in maternal care during the weaning phase. In the wild Barbary macaque, collective care of young by males, including a great deal of carrying and comforting of individuals, occurs. Adult males, subadult males, and subadult and nonmaternal females are all involved in this activity, which is also associated with an advanced agonistic buffering system described later in the chapter. In some species infant adoption by males has been reported—not surprisingly perhaps among baboons, macaques, and chimpanzees, which have been studied sufficiently as to allow these data to come to light. Young primates without some form of adult care would seem unlikely to reach maturity or to develop the emotional stability upon which their own reproductive behavior depends. The fact that adoption can occur is a strong piece of evidence in favor of systems of caring over and above those involved in direct parenthood.

These examples of generalized and collective caring suggest that male care is a much more widespread phenomenon (at least in primates) than had formerly been supposed. The implications are that, apart from behavior that can be correctly described as parental, the male invests a good deal of time and energy in apparently altruistic caring of infants other than those directly his own. Care between relatives other than parent–child may be elaborate, as where young males excluded from one troop in the rhesus Cayo Santiago population were aided in joining another group by the protective adoption behavior of the elder brothers who had preceded them there. It seems clear that as longitudinal studies develop and the genealogical relationships among troop members become better known in study populations we are likely to find extensive networks of care-giving and -receiving behaviors that will probably prove to be characteristic of closely related lineages of a clan type.

In viewing the scattered information available and the direction in which research findings seem to be leading, we may well ask what, in terms of evolutionary theory, the male gains from this behavior. A valid approach would be in terms of the evolution of so-called altruism through kin selection (Hamilton, 1963). The essential idea here is that there are situations where an individual is more likely to transmit a proportion of its genetic complement to the next generation if it assists kin than if it exclusively occupies itself with its own progeny. Among birds there are certain communal arrangements wherein a whole clan of relatives of at least two generations may assist one another in incubating, feeding, or otherwise tending the young. Often there is evidence suggesting that without such help single females or pairs working alone would not be successful.

We may argue that among primates collective care is especially pronounced where paternity is in question. Thus in species where the reproductive male lives in one-male groups, his collective care is mainly directed to his own offspring, albeit of different females. In multimale troops, however, the dominant males mate with a range of adult females and have, in a sense, fathered the young of the season in common. Their defense and care-giving turns out also to involve much collaboration. Since these males are likely to be quite close kin, this cooperation amounts to a mutual insurance that at least some of the collectively held genetic material will get through to the next generation. Such behavior is possibly most apparent in well-integrated groups with relatively clear status and role differentiation. The males' behavior here becomes a virtual imposition of norms on the local population. Essentially, in macaque troops and among baboons, the adult males control the level of affect by policing quarrels so that social disruption is minimized. In a study by Vandenbergh (1967), it was shown that macaques of a group introduced to a small island off Puerto Rico, lacking a matrilineal genealogy and dominant males, were socially very unstable, and few young were successfully reared.

Male reproductive strategies in primates, and probably in other highly sociable mammals such as carnivores, may thus contain a caring component that tends to ensure the survival and rearing to maturity of not only the mammal's own young but those of its kin as well. The situation is, however, a rather odd one as, where a choice is available, we would predict that a male would give greater care to its own offspring than to those of relatives. So far there appear to be no sufficiently critical studies to indicate whether this is so and in what way different individuals balance their orientation of caring behavior among possible recipients. We might expect, for example, that the more dominant adult male reproductives in a multimale troop of baboons would be the troop members most concerned in caring, since they are probably the progenitors of most of the young. Lower rankers would be concerned with the displacement of the dominants from their reproductive hegemony and would be indifferent to the fate of the dominants' young, except when they were also their own close relatives. Nonreproductive or subordinate males are likely to seek ways of improving their social position, not only because this may give them easier access to desirable commodities but also because with access to females they may improve their reproductive success.

It is at this point that it seems useful to reconsider the phenomenon of agonistic buffering in primate group life. Deag and Crook (1971) as well as Itani (1959) noted that in the macaque groups they were studying subordinate males would often utilize care-giving behavior to young as a signal to a third party in a triad—usually a dominant male. The young infant, especially if it is still in its natal coat, seems to elicit behavior from the dominant animal in which the aggression of the latter, if present, is rapidly reduced and is followed by its initiation of some friendly overture. Evidence for this comes from macaques, langurs, and several species of baboon, as well as from the gelada. This inhibition on aggression appears to be due to a motivational antagonism between tendencies that eventuate in assertive–aggressive interactions and those eventuating in care-giving or grooming behavior. In the context of selection's favoring secure social environments for developing young through the emergence of more or less altruistic behavior, such an inhibition makes good sense. The trick is that monkeys involved in social climbing can turn the response to their advantage.

While work on this topic has hardly developed from initial speculations, it is clear that the presentation of a baby by a subordinate Barbary macaque to a dominant and potentially aggressive one (Deag & Crook, 1971) does reduce the likelihood of an agonistic interaction. The behavior, which occurs in a predictable ritualized exchange of formalities including baby bottom sniffing, appears to buffer social relations between males. Indeed, outright aggression appears rare in this species. When aggression does occur, one male often dashes for an infant as a means of abating the wrath of his opponent.

Itani (1959) expressed the view that males that related closely with young of dominant females and used them in agonistic buffering could themselves rise in social status. While this viewpoint requires further detailed examination in the field, it does seem a likely function for some of the behaviors involved. Social subterfuge is commonly employed among sociable primates to bring about patterns of relating without too much antagonism. The way in which young male hamadryas baboons and geladas acquire reproductive groups illustrates this very well. One method now well researched by Kummer (1968) and by Robin Dunbar (1973) is for young males to attach themselves to existing groups. While the adult males present usually resist the low-key approaches made by the would-be entrants, the latter usually succeed by persistently associating with the group. In both species, although in different ways, the young males relate increasingly to very young females, which eventually become the sexually mature nucleus of the young male's own group. The way in which the young male hamadryas attracts the young females' attention is to act in many ways as a mother substitute at around the time when the mothers tend to be decreasingly concerned with these young. The young male provides protection in the event of small quarrels and also contact comfort through grooming and other activities. These males thus make use of their care-giving potential to establish bonded relations with members of the opposite sex. Only on rarer occasions do younger males challenge the group males directly for sexual access to adult females. In the gelada, at least, this fate is reserved for aging males. Dunbar's research shows that, in the fights that ensue, the older male begins to lose only as his females begin to transfer their allegiance to the younger animal. We need further accounts of these tremendous battles, which are not often witnessed and which shed much light on the sex relationships in this species.

The conflict between behavior eventuating in care for kin and that exclusively concerned with a male's own offspring sometimes erupts in explosive events of a tragic character. There are now several accounts showing that, among langurs in particular, a maturing male may attack a male reproductive group with ferocious intensity and, after defeating the owner, slaughter his babies. The matings with the females that follow then establish that the young of the reconstituted group are the progeny of the male present within it. Were the male not to kill the offspring of the former owner, considerable time would have to pass before he could get his own genes into generational transmission.

Nearly all the elaborate behaviors we have been describing are still poorly known and need much further study. Their occurrence does, however, open new vistas in socioethological research. So far as the males of these highly sociable primates are concerned, we can see three closely entwined strategies at work, all of which are ultimately aimed at furthering genetic transmission:

1. The development of cooperative caring behavior, which increases the likelihood of the survival of a male's own young.
2. The development of collective caring, irrespective of paternity but likely to be directed to close kin. Such behaviors appear not only in reproductives but also in other relationships, such as those between brothers.
3. The development of social subterfuge whereby the aggression-inhibiting aspects of caring behavior or the caring relationship itself is utilized to modify the behavior of a third animal. The effect seems likely to lead to changes in status which enhance the probability of a male's reproduction. (Such behavior also allows the male hamadryas baboon to establish a harem of nonreproductive females, as it were, under the nose of their father.)

All these behaviors require considerable subtlety in the use of social skills, which probably demands a long period of motivational maturation in childhood. Indeed, the primary selection pressures in these animals are probably social. They strongly favor the emergence of means whereby individuals not only maintain appropriate relations within a complex social system but also acquire the art of entering into intersexual relationships, probably against the social constraints generated by the hegemony of older incumbents; those behavioral legislators whose anxious defense of the status quo reflects their motivation to remain reproductive as long as possible.

The outcome of these masculine social games is far from being uninfluenced by the females themselves. Much stress has been laid on the domineering character of the male hamadryas baboon, who rules his females with the neck bite. Things are far from comparable in either the patas monkey or the gelada, although both have social organizations whose structure is considerably similar to that of the hamadryas. In both patas and gelada the male is a somewhat peripheral member of the reproductive group, except during the early phases of the unit's establishment. The real political center of the mature gelada group, as my colleague Robin Dunbar is at present showing, resides in the close relationships between certain of the adult females. Ultimately it is they who determine directions of march and foraging, it is they who may call up the males' aid in conflict situations but who ultimately determine their outcome, and it is they who, by changing loyalties in relationship, may eventually bring about the demise of an aging harem owner. The patas male is even more peripheral since, as Hall (1966) showed some years ago, he spends much time away from the group in a tree looking out as kind of watchdog. Recent work by Struhsaker and Gartlan (1970) suggests that this preoccupation expresses not so much a concern with predators as with the whereabouts of other patas groups—particularly all-male groups from which harassment might come. Again, here it seems likely that females exercise a great deal of choice in determining which males occupy the reproductive role.

The female's strategies in these situations are again presumably a function of the biological advantages to be gained in ensuring genetic transmission. We have already suggested that this process will be facilitated if she mates with a male of strong genetic constitution. The male's characteristics of display and vigor may well exercise a major influence on her choice and also in determining her shifts in relating from one male to another. In addition, females of the more sociable primates may be looking for characteristics that indicate a strong caring capacity in the male. A caring male is perhaps especially likely to attract females and perhaps to run a highly productive reproductive group. In addition, she probably looks for a male that will tend to reduce social stress in the social unit, both through modifying participation in within-unit quarrels and in maintaining group boundaries against disturbance by other males. A male's vigor in group maintenance may be well indicated by the extent and manner of protected threat behavior which he offers young females when the latter get into conflicts with third parties early on in the bond-forming stages of their relationship.

We have been looking at some of the most complex types of intersexual interaction so far uncovered in primate research. We still have little knowledge as to why the relative influence of one sex on the other in the social dynamics of the reproductive group differs so much between species. The differences can be treated as functions of contrasting phylogenetic history but, like the old appeal to instinct, this gets us no further. The answer can only gradually take shape in extensive longitudinal studies of population and social dynamics of contrasting species in which the full complexity of these behaviors is examined in relation to their social and ecological functions.

Conclusions

In conclusion, it is perhaps at least permissible to ask what all these contemporary findings mean for our own species. My own view now is that the complexities at the nonhuman primate level are sufficient in themselves and that, until we have much greater insight into them, any generalizations to our own species are likely to be no more than a speculative game. It seems reasonable to argue that strategies similar to those we have been discussing probably played important roles in the emergence of intersexual relationships of protohominid primates and that an understanding of our biological heritage may emerge from primate studies of this kind. Even so, I believe that such an understanding provides no more than a kind of educational back-cloth—a reference literature for university courses on human evolution—and that it cannot begin to touch upon the existential issues that are the central focus of living human relations.

Ethologists have not yet begun to grasp at the crucial characteristic of human beings that tends to invalidate all but the most cautious inferences from other animals to man. Surprisingly, too, psychologists have also often failed to recognize the point I am about to make. As a species man possesses the characteristic of self-reference. This means that not only do we describe ourselves to ourselves and create identity constructs but we can also face the issue of whether we want to alter those constructs and become another kind of person. Different phases of human history have thrown up different gender identity constructs. Our own period is not only one in which gender identities are in the process of major changes but, in addition, it is one in which we are highly aware and articulate about our views on the nature and desirability of that process. Many nonhuman primates show complexity and flexibility in their gender relations—our species does so in a transcending degree. Whatever the constraints of a biological nature we may have inherited, an overruling human characteristic remains behavioral flexibility and the capacity for generating relationships of many kinds (Crook, 1973).

I feel there can be only one conclusion. If ethologists are truly interested in an ethology of human relating, we have to drop the pretense that, given present knowledge, evolutionary arguments are any more than a game for the psychologically literate. Instead we should look directly in the place where the answer lies—the proper study of mankind. The proper study of sex relating in our species is to seek direct understanding of the way men and women interact with one another in love and conflict, in rejection and in mutual comfort, in their antagonisms and in their complementarity.

References

Buechner, H. K. Territorial behavior in Uganda kob. *Science*, 1961, *133*, 689–699.

Campbell, B. (Ed.), *Sexual selection and the descent of man: 1871–1971*. Chicago: Aldine, 1972.

Crook, J. H. The socio-ecology of primates. In J. Crook (Ed.), *Social behaviour in birds and mammals*. New York: Academic Press, 1970.

Crook, J. H. Sexual selection, dimorphism and social organisation in the primates, 1972. In B. Campbell (Ed.), *Sexual selection and the descent of man: 1871–1971*. Chicago: Aldine, 1972.

Crook, J. H. Darwinism and the sexual politics of primates. In *L'origine dell' uomo*. Rome: Accademia Nazionale dei Lincei, 1973. Also in *Social Science Information 12(3)*, 7–28.

Crook, J. H., Ellis, J., & Goss-Custard, J. Mammalian social systems: Structure and function, *Animal Behaviour*, 1976, *24*, 261–274.

Deag, J., & Crook, J. H. Social behavior and 'agonistic buffering' in the wild Barbary macaque, *Macaca sylvana L. Folia Primatologica*, 1971, *15*, 183–200.

Dunbar, R. *The social dynamics of the gelada baboon, Theropithecus gelada*. Ph.D. thesis, *Bibliotheca Primatologica*, Bristol: University Library, 1973.

Eisenberg, J. F. The social organisation of mammals. *Handbuch der Zoologie*, 1965, *10(7)*, 1–92.

Estes, R. D. Behavior and life history of the wildebeeste (*Connochaetes taurinus. Burchell*). *Nature*, 1966, *212*, 999–1000.

Goss-Custard, J. D., Dunbar, R. I. M., & Aldrich-Blake, F. P. G. Survival, mating and rearing strategies in the evolution of primate social structure. *Folia Primatologica*, 1972, *17*, 1–19.

Hall, R. Behaviour and ecology of the wild patas monkey *Erythrocebus patas* in Uganda. *Journal of Zoology*, 1966, *198*, 15–87.

Hamilton, W. D. The evolution of altruistic behaviour. *American Naturalist*, 1963, *97*, 354–356.

Itani, J. Paternal care in the wild Japanese monkey, *Macaca fuscata. Primates*, 1959, *2*, 61–93.

Kleiman, D. G., & Eisenberg, J. F. Comparisons of canid and felid social systems from an evolutionary perspective. *Animal Behaviour*, 1973, *21*, 637–659.

Kummer, H. Social organisation of hamadryas baboons. *Bibliotheca Primatologica*. Vol. 6, 1968.

Lack, D. *The natural regulation of animal numbers*. Oxford University Press, 1954.

Mitchell, G., Redican, W. K. & Gomber, J. Males can raise babies. *Psychology Today*, 1974, *April*.

Orians, G. On the evolution of mating systems in birds and mammals. *American Naturalist*, 1969, *103*, 589–603.

Schaller, G. B. *The serengeti lion*. Chicago: University of Chicago Press, 1972.

Struhsaker, T., & Gartlan, J. S. Observations on the behaviour and ecology of the patas monkey (*Erythrocebus patas*) in the Waza Reserve, Cameroon. *Journal of Zoology*, 1970, *161*, 49–63.

Trivers, R. L. Parental investment and sexual selection, 1972. In B. Campbell (Ed.), *Sexual selection and the descent of man: 1871–1971*. Chicago: Aldine, 1972.

Vandenbergh, J. G. The development of social structure in free ranging rhesus monkeys. *Behaviour*, 1967, *29*, 179–194.

3

The Evolution of the Reproductive Unit in the Class Mammalia

John F. Eisenberg

I heard Daniel Lehrman lecture for the first time in 1957. The subject was ring doves and the hypotheses concerned the interrelationship of physical objects (nests and eggs), behavior (courtship), and induction and phasing of hormonal secretions. It was and is an elegant story. Although I have studied mammalian reproduction and social behavior from the perspective of a phyleticist, the paradigm proposed by Lehrman has never been forgotten. Some years later, in 1970, I was invited to present a lecture on the reproduction patterns of tenrecoid insectivores at the Institute of Animal Behavior. Lehrman's interest, penetrating questions, and courtesy made me realize that even though we worked from widely different points of view, a zoologist such as myself had much to gain and perhaps something to offer to the active, experimental group at Rutgers. I am both touched and honored to have been invited to contribute to this first symposium honoring Daniel Lehrman, phyleticist at heart, who founded the active, experimentalist group which hosted us today.

Introduction

Anyone engaged in the process of defining a term is limited by the state of knowledge at the time of formulating the definition as

John F. Eisenberg · National Zoological Park, Smithsonian Institution, Washington, D.C.

well as by those philosophical biases the individual may bring to the material. Perhaps one of the most ambitious efforts in animal sociology was made by Deegener (1918), who exhaustively named social units according to recognizable functional and compositional categories. His total reached some 58 social types, not including forms of aggregation or temporary association. All of his terms were developed *de novo* by the appropriate combination of Greek prefixes and suffixes. His terminology was demanding and rather esoteric. In spite of its rigor, I know of only Krumbiegel (1953, 1954), who made extensive use of this scheme for mammals.

Most investigators of animal social behavior attempt to build their definitions of social units around the very important fact that different species show different dispositions in space. We then arrive at a classification based on spatial criteria. Figure 1 refers to a tripartite classification: colonial, communal, and dispersed. Spacing is maintained through the use of threat display and negatively reinforced by agonistic encounters. Presumably such spacing mechanisms involve the partitioning of resources within an environment and vary from species to species as a result of a compromise between (a) the ability to defend and (b) the utility of discrete defense of restricted resources (Brown, 1964). This assumes a center of activity for each individual and, in the case of the dispersed system, an individual's activity areas are for its exclusive use. In a colonial system, exclusive use areas are still present, but the memebers of each colony are clumped into discrete groupings. In a communal system, all individuals have potentially equal access to resources within the area that the group occupies. This sort of simple classification is useful for species that have a fixed location in space.

Referring to Figure 1, the colonial system might represent the territories of adult, reproducing yellow-bellied marmot females (*Marmota flaviventris*) (Armitage, 1962). The communal diagram may represent the coteries composing a prairie dog town, such as those described by King (1955) for *Cynomys ludovicianus*. The dispersed system would exemplify the foraging territories of the adult red squirrel (*Tamiasciurus*) described by Smith (1968).

In the case of the red squirrel population studied by Smith, males and females had individually distinct, defended territories. Males only entered females' areas at the time of reproduction. This is an exception rather than the rule, since in most cases a male's home

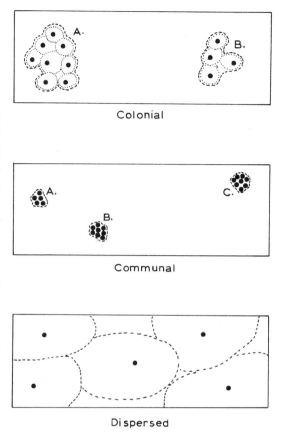

Colonial

Communal

Dispersed

Figure 1. Three possible schemes illustrating animal spacing patterns. Black dots equal individuals. Dotted lines equal boundaries of individual or group living areas. Colonial: Individuals are grouped into distinct clumps but each individual has access to an exclusive area. A and B are two colonies. Communal: Several individuals share a home range in common. They may be discretely organized (A, B, C) or clumped into colonies; thus, a communal system may become a special case of a colonial system. Dispersed: Individuals have exclusive use areas which may be true territories under special circumstances.

range may overlap a female's and he may have access to many parts of it, although perhaps not access to the natal nest at the time of parturition.

This brings us, then, to a consideration of the reproductive unit. The reproductive unit can be considered a social structure organized to replicate itself through births. Spacing mechanisms may be weakly developed in the case of communal species or strongly manifested in the case of species exhibiting a dispersed system. If an area is defended for exclusive use, it is referred to as a territory. If, on the other hand, the foraging area of an individual is shared with others, it is generally referred to as an individual's home range (Burt, 1943). It should be understood, however, that even with communal use patterns or considerable overlap in foraging areas, each individual in an organized social group may show an individual distance, which means that when it is in a given part of its home range a neighbor can only approach to a certain distance without being threatened or attacked outright (Hediger, 1952). As a result of negative reinforcement, such agonistic bouts can lead to the formation of a hierarchy of individuals that use the same space in common yet exhibit differential access to resources with the space. Even if the individual organisms move about and have no fixed site attachment, individual distance may be shown and hierarchies may be formed. Members of a given social group can have an internal hierarchy and mutually defend themselves against threats from conspecifics in areas of overlap (Ewer, 1968).

Fisler (1969) made a significant advance in classifying social units, approaching the problem by cleanly separating those organizational systems based on individual effort from those based on group efforts. Under each system, he then subdivided according to a basic recognition of two functional classes of behavior: (a) site attachment and (b) capacity to express aggressive behavior. Both types of behavior could vary in step as one passed from an extreme case of high site attachment and low interindividual aggression. At this latter point, group aggressive systems could become an emergent phenomenon. In his system, he makes due allowance for seasonal variation or phases.

In my publication of 1966, I recognized two broad types of social systems. In one type, the individuals are relatively dispersed in space and come into intimate contact only when a male and female mate or during that time when the female is preoccupied in the initial care and rearing of the young. This so-called solitary or asocial system has run into severe semantic difficulties and has been criticized by Anderson (1970). Perhaps the choice of the word *solitary* is unfortunate. Leyhausen (1965a) wrote a very provocative paper titled "The Communal Organization of Solitary Mammals," in which he expounded upon the paradox. Yet I have chosen to retain the term *solitary* for convenience, fully recognizing that, in order to remain apart from one another, adults must know the position of their neighbors, and thus some form of communication or monitoring of conspecifics must take place. I have justified this in arguments elsewhere (Eisenberg, Muckenhirn, & Rudran, 1972; see also Seidensticker, Hornocker, Wiles, & Messick, 1973, p. 54).

In contrast to this, we may have rearing systems composed of several adult animals. Such groupings, when mobile and moving over a wide area that is not uniformly defended, show a great deal of cohesion. A discrete structure can be observed either at certain times or throughout the reproductive cycle. In my own efforts at a synthesis (Eisenberg, 1966, 1967), I duly noted the phasic nature, often tied to reproduction, of social units, and I then developed a scheme to include four major forms of social structure, all ultimately based on variations in the extent of parental care. The evolution of a cohesive social grouping, which exhibits forms of protocooperation, is almost always tied to the development of extended families, and thus the group structure has a kinship network as its basis. Other forms of sociality, which are often more ephemeral, do not derive from parental care units.

Using the sex and age composition of such groupings as criteria, I proposed a simple classification: (1) The matriarchy or extended mother family consists of females, in part related by descent, which form a unit for the mutual rearing of their progeny. This is one of the more common types of cohesive social groupings shown in those species of Mammalia which form organizations composed of several adults. (2) Some social systems involve continuous male association but only indirect participation by the male in the rearing of the young. The male, through his activities, aggressively keeps other males out of the living space, thus increasing the carrying capacity of the area in which his females are located. This, then, would be a unimale rearing system generally exhibiting polygyny (e.g., zebra or horse, Klingle, 1972). (3) A rare symptom in the Class Mammalia includes the participation of the male with the female in the initial rearing phase of the young. This, then, is a true parental family. The role of the male is varied and may involve provisioning, such as in many species of the genus *Canis* (Kleiman & Eisenberg, 1973), or carrying of the young, shown uniformly in the New World primate family, Callithricidae. (4) One of the most infrequently evolved social systems in the Class Mammalia is that which includes a relatively permanent, cohesive grouping of several adult reproducing females attended phasically or continuously by several adult males potentially having partial access to the sexually mature females. Such systems have evolved convergently in the lion (Schaller, 1972), some primates (Crook & Gartlan, 1966), some Cetaceans, and a few ungulates (Eisenberg, 1966).

As McBride (1964) points out, one must not think of vertebrate social systems as static entities, either in space or in time. With a simplified diagram (Eisenberg, 1966, p. 15), I tried to indicate that regardless of the form of dispersion in space of the species in question, at least three critical phases occurred in the development of the next generation: (1) the time during which the adult male and female pair and mate, (2) the time of parturition and early maternal care, followed by (3) the time during which the littermate group moves about with the female prior to dispersal. Even in a simple system, where the young disperse early and the male does not consort with the female, these three phases are of paramount importance. Obviously, the phasing of social tolerance, even in the most asocial of species, permits pairing and the constitution of, at the minimum, a mother family. Thus, basically, there will be a period of pair tolerance. The duration of the pair association may in fact be brief since the unique method of neonatal nutrition in the Mammalia makes permissible a maximum parental investment by the reproducing female.

Recently, Anderson (1970) has criticized Fisler and myself, rightly pointing out that both of us have defined a type of social system referred to as "asocial" or "solitary" (Eisenberg, 1966, 1967) or "exclusive territorial" (Fisler, 1969). Although both of us recognize the universality of the parental care unit (in most cases a "mother family"), Anderson focuses on the reality of the deme as the fundamental social unit. He suggests that, in most populations of mammals regardless of the spacing system, the fundamental organization consists of a set or several sets of parents (monogamous or polygynous) with offspring of various ages. Such demes are often divisible into functionally different classes: (a) a stable reproducing unit from which offspring disperse and (b) newly formed units which form from dispersing individuals and serve as colonizers. Not only does such a demic view of populations have much heuristic value for the ecologist, but it is also a fundamental structural concept for understanding the role of kin selection in the differential survival of genetic traits (see Wilson, 1973). Yet, the population structures described by Anderson are not incompatible with the recognition of different degrees of social tolerance as outlined by Fisler and myself, for, undeniably, mammalian species vary with respect to their tolerance for conspecifics and in the forms of spacing mechanisms which they show (Eisenberg, 1967; Crook & Gartlan, 1966).

The Reproductive Unit as a Device for Self-Replication

It is essential to consider survival strategies when attempting to understand the evolutionary histories of mating systems. Survivorship of individual genotypes (or portions of genotypes) is ensured only if the reproducing individual replicates itself through the creation of a viable offspring reproducing in the next generation. Thus, in a species exhibiting a minimum of social tendencies, the reproductive strategy of the male will be to select a female for insemination with the goal of leaving behind his genotype as expressed in the joint progeny reared by the female. This assumes minimal parental investment on the part of the male. Thus, his choice of a female will be critical if he is monogamous and breeds only once in his lifetime. On the other hand, the male may opt for a strategy of inseminating as many females as possible, thus increasing his chances for leaving behind progeny (Orians, 1969).

The female, on the other hand, will be concerned with rearing a litter and producing at least one surviving offspring in the next generation. If the male does not participate in parental care, then her selection of a male may not be nearly as critical. On the other hand, if the male does participate to some extent, either in parental care or in enhancing by his efforts her capacity to rear the litter, then her selection of a male to mate with becomes more critical. In any event, the female will have to have access to resources such as food and water in sufficient quantities to guarantee that she will have the economic means at hand to rear her litter. This is especially critical in species that breed during only one season. In a long-lived species with several chances at reproduction in its life history, such availability of resources may be less critical in any one season but, overall, may have the same impact.

Examination of closely related species, such as rodents of the genus *Marmota*, indicates interesting trends in the evolution of mating systems. For example, in the yellow-bellied marmot, *M. flaviventris*, the adult females are somewhat aggressive toward one another, thus spacing themselves out so that sufficient foraging area is available to each of them in order to guarantee sufficient sustenance to see them through their lactation and rearing phase. In an especially rich habitat with a high carrying capacity, females may in effect defend a rather small foraging area. The net result is that a given male can defend with very little effort several females and thus approximate a polygynous mating system. In areas of poor carrying capacity with a large home range per female, a male may find it economically impractical to defend an area that encompasses more than one female. In this case, the system appears monogamous. Yet the potentiality for expressing a polygynous mating system is there, but in this case its expression is a function of habitat (see Armitage, 1962; Downhower & Armitage, 1971). In this species, there is little direct participation on the part of the male in the rearing of the young; nevertheless, through his own activities, he ensures that a number of adult males do not occupy the same area as the female. Thus, he enhances, through his territorial activities, the reproductive success of a female within his range. This sort of indirect benefit that a female can derive through the defensive presence of a male is widespread in mammalian species that do not show a permanent family unit.

Lest one think that this system typifies reproduction in the genus *Marmota*, however, it should be pointed out that *M. monax*, the woodchuck, shows even less structure in its social unit. The young woodchucks mature very rapidly during the spring of their birth and, through aggressive activities between the adults and themselves, they disperse to establish independent foraging areas prior to their autumnal hibernation (Bronson, 1964). On the other hand, *M. flaviventris* does not show such a quickened rhythm in dispersal of littermates, and the young remain with the female, feeding in her core area and entering hibernation, only to disperse in the following spring.

Barash (1973) has demonstrated yet a more complex social structure in the Olympic marmot, *M. olympus*, where the young of the year do not disperse until their second or third year. Thus, in the Olympic marmot, there is an extended family occupying a defended area with an adult male, one or two adult females, and young of two generations. This unit shows the beginnings of some cooperative behavior. Several individuals may work on the same burrow system. Vocalizations which serve to alert the colony of the presence of potential predators may be given by all colony members, thus increasing the potential for appropriate antipredator behavior by any individual within hearing range, etc. Barash relates the differences in social structure among these marmot species to differences in the length of the plant growing seasons. For *M. monax*, the growing season is long and the young of the year appear to be able to obtain sufficient food to allow them to reach adult size by autumn; thus they can breed in the following year. The yellow-bellied marmot, occurring at higher elevations, has a shorter growing season, takes longer to mature, and does not breed until two years after its birth. The Olympic marmot, restricted to alpine meadows, takes even a year longer to mature.

This brings up an interesting correlation, then: that extended family units very often involve a slow maturation rate, a greater potential longevity for individuals, and a lower reproductive capacity per individual female. This permits some degree of generation overlap and some emergent survivorship benefits deriving to the young generations from such an association with the adults. Thus, in the case of *M. olympus*, the functional reproductive unit is no longer the mother holding her territory for a given breeding season but the whole parental-offspring complex mutually sharing a territory through several breeding seasons.

The Evolution of Reproductive Systems

We must now explore the question of evolutionary history. In order to do this, we have to make some sort of educated guess concerning the ecological and behavioral baseline from which all contemporary mammalian species derive. Since we cannot study the behavior of fossil forms, we must make inferences concerning their behavior based upon their morphology and upon our interpretation of the mode of habitat exploitation that the extinct species showed. Thus, it is essential to study those species alive today which exhibit a conservative morphology, resembling in their brains, sense organs, and bodies the structure of the early mammals. Such conservative species that occupy niches which we believe to be similar to those occupied by mammals at the Cretaceous-Paleocene boundary will serve as a behavioral baseline (see also Jerison, 1973).

If we are to look for evolutionary trends, let us first formulate a working definition of *conservative morphology*, as well as a definition of *conservative niche*, and then examine the trends in the evolution of courtship, copulation, gestation, and parental care. The exact steps in my argument for the definition of a conservative niche are outlined in a previous publication (Eisenberg, 1975); thus, I will only sketch the sequence here. Suffice it to say that the Class Mammalia embraces a morphological grade with a polyphyletic origin (i.e., the three living orders attained the grade of mammalian morphological organization independently). It is safe to assume that, as these early premammals adapted, they did so in response to a nocturnal, forested niche. They probably climbed reasonably well and relied on chemical and auditory inputs for distance reception. Auditory input was especially important. The eye was probably used for gross differentiation of distant objects. Early on, the mammals evolved homiothermy, and the female became specialized for producing nourishment for the neonate with the evolution of mammary glands.

It seems probable that the trend away from strict oviparity with large-yolked eggs toward the laying of small-yolked eggs together with the development of mammary glands was evolved in a parallel fashion by both the ancestors of the Monotremes and the ancestors of the Pantotheres. The latter were ultimately ancestral in the Marsupialia and the Eutheria. In these latter two taxa, viviparity was evolved. The marsupials have yolky eggs and typically show the formation of a yolk sac placenta or, in the Peramelidae, a temporary chorioallantoic placenta. Intrauterine development is brief. The eutherians evolved the embryonic trophoblast, chorioallantoic placentation, and a longer gestation period. We can discern in both the marsupials and eutherians advanced and conservative characters of reproduction which are parallel or convergent. In both groups we find that the specialized or "advanced" forms show: (1) a reduced number of ova shed at the time of ovulation, (2) a reduced number of young, (3) an increase in the weight of the neonate, and (4) a lengthening of the gestation period (Portman, 1965; Sharman, 1965). Of course, in all the Marsupialia, the neonate is much less developed at birth than is the case in even the most conservative eutherians. It is fair to say that primitively the Eutheria probably produced several rather altricial young which were initially brooded and suckled by the mother in a nest. On the other hand, in the Marsupialia, the conservative forms produced a number of extremely altricial young which by means of their own motility transferred to a teat area for attachment and nourishment. In most of the Marsupialia, the teat area is enclosed in a pouch.

Some eutherians and marsupials today are undoubtedly occupying niches that are similar to those occupied by ancestral forms at the

Paleocene–Eocene boundary. We may speak of these forms as occupying conservative niches and showing a conservative morphology. The following trends are exhibited by such forms. Conservative mammals primarily feed on energy-rich foodstuffs readily digestible, such as invertebrates, small vertebrates, or fruits. Feeding on plant bodies directly, with the capacity to degrade cellulose, aided by microbial symbionts, is a much later evolutionary adaptation. Although morphologically conservative forms may show reduced metabolic rates when compared with more so-called advanced species, they nevertheless show homiothermy even though they may have evolved facultative hypothermia as a means of passing through periods of food scarcity. Because of their food requirements, these conservative species tend to range over a relatively large area to procure their food. Most of these forms are clearly nocturnal, although the eye may not be necessarily reduced in size. Some frontal vision is possible, and their climbing ability is well developed. Audition is extensively employed in the reception of signals from distant objects. Vision is only secondarily useful in resolving objects in space.

Whether we consider either a eutherian or a marsupial, it is fair to say that the precursors were probably relatively small (50- to 150-gram animals), insectivorous or omnivorous, produced large litters of relatively altricial young and were predisposed for individual foraging patterns and, very probably, a dispersed social organization. For this reason, I choose to consider the contemporary *Marmosa robinsoni*, a marsupial, and other members of the insectivore family, Tenrecidae (*Microgale dobsoni*, *M. talazaci*, and *Hemicentetes semispinosus*), as forms which resemble the early mammals in their foraging patterns and morphology. The living Monotremes (the echidnas and platypus) represent specialized derivatives of an earlier radiation.

This is not to imply that there cannot be specialization within a conservative morphological framework; indeed, many would argue that *H. semispinosus* is highly specialized, which it certainly is. It has specialized for feeding on earthworms with concomitant modifications in its dentition, skull, and forepaws; it has lost its tail, its eye is vastly reduced, and instead of fur it has a spinescent coat and a complicated antipredator defense mechanism. Such specializations offer instructive insight into the limits of specialization within a conservative morphology (Eisenberg & Gould, 1970).

Courtship and Mating

In morphologically conservative species, a given male and female may show overlap in home range and yet show very limited contact except at the time of mating. Parental care generally falls entirely to the female. Nevertheless, the same male and female may mate during consecutive seasons as the result of the proximity of their home ranges and as a result of their own agonistic tendencies toward conspecifics of the same sex. When a male and female establish contact for the purpose of mating, there are preliminary interactions which might be relatively stereotyped in either configuration or sequencing. Such predictable sequences are thought to involve information exchange and may include aspects of chemical, auditory, visual, and tactile input.

Let us then review the trends in mating behavior shown by such morphologically conservative mammals. It will be noted that the production of sound (clicks during the courtship of *Marmosa*, piffs during the courtship of *Tenrec*), olfaction (e.g., nasogenital sniffing and marking), and tactile input (e.g., licking and touching) form prominent aspects of the initial courtship behavior (Eisenberg & Gould, 1970). Visual display is rudimentary at best (see Gould, 1969).

Chemical mediators are strongly implicated in signaling receptivity by the female, and the odors from the male often induce receptivity on the part of the female (for a review see Eisenberg & Kleiman, 1972). The work of Devor and Murphy (1973) and Murphy (1973) on *Mesocricetus auratus* demonstrates that the female produces a substance in her vaginal secretions which actually triggers mounting and copulation on the part of the male. If the olfactory nerves are severed in the male hamster, he will be unable to complete copulation. Thus, in the case of the hamster, odors from the female act on the male almost as classical releasers.

The "priming" effect of odors, sounds, and touch must not be underestimated regardless of whether the female is an induced or spontaneous ovulator. Of course, the distinction between induced and spontaneous ovulation is arbitrary at best. Actually, there appears to be a continuum between, on the one hand, a strict induced ovulator, such as the rabbit, *Oryctolagus*, and, on the other hand, the spontaneously ovulating laboratory mouse, *Mus musculus* (Richmond & Conaway, 1969).

There have been too few detailed investigations on mammalian courtship to illuminate the effectiveness of various sensory inputs on the induction of receptivity on the part of the female or in the stimulation of the male. It would appear that morphologically conservative eutherian mammals employ chemical signals in the induction of mating, and it would further appear that the females require considerable stimulation to induce ovulation (Eisenberg & Gould, 1970).

One correlation that emerges from an inspection of courtship and copulation of "primitive" mammals is that once intromission has taken place it tends to be long in duration. Morphologically conservative marsupials, as well as eutherians, tend to show a prolonged intromission time (see Table I). Such intromission times seem to correlate with the induction of ovulation and the fact that the female is primed to receive a certain amount of sustained tactile input during copulation. An advancement away from this process of sustained intromission in other lines of mammals seems to have involved: (1) the use of many short intromissions to achieve maximum stimulation of the female, or (2) highly synchronized ovulation induction mechanisms which require a brief copulation with subsequent ovulation on the part of the female, or (3) brief intromission times with maximum stimulation of the female resulting from the evolution of special spines and plates on the penis of the male (see Kleiman, 1974a, for a review of this phenomenon in the caviomorph rodents). Such trends in copulatory behavior have occurred convergently and in parallel fashion across the mammalian orders.

Thus, it would appear that one phylogenetic trend has been to evolve signals which more closely coordinate the timing of copulation and ovulation in the female. The net result may be to reduce actual courtship time in some cases and, in other cases, to reduce the amount of time spent in copulation. Which solution has been selected for, when various species are compared, depends on a great many environmental factors, which are poorly understood.

Aside from these changes in copulation patterns, the basic interaction during courtship leading up to copulation has remained remarkably conservative. Even in such highly specialized forms as the elephant (*Elephas maximus*), tactile and olfactory inputs are primary

Table I. Some Intromission Times for Selected Mammals

Taxon	Range or average intromission duration	Average number intromissions until ejaculation	Author
Marsupialia			
Didelphidae:			
Didelphis marsupialis	~28 min	1	Reynolds
Marmosa mitis	50 min–5 hr	1	Barnes & Barthold
Dasyuridae:			
Antechinus stuarti	up to 9 hr	1	Collins, Ewer
Sminthopsis Crassicaudata	2 hr–11 hr	1	Ewer
Macropodidae:			
Macropus cangarou[a]	20 min–50 min	1	Sharman, Calaby, & Poole
Megaleia rufa[a]	10 min–25 min	1	Sharman *et al.*
Peramelidae:			
Perameles nasuta[a]	2–4 sec	Several	Stodart
Insectivora			
Tenrecidae:			
Hemicentetes semispinosus	15–25 min	1	Eisenberg & Gould
Setifer setosus	25–<70 min	1	Eisenberg & Gould
Echinops telfairi	17+ min	1	Eisenberg & Gould
Soricidae:			
Blarina brevicauda	3–5 min		Pearson
Suncus murinus[a]	<15 sec	av. 38	Dryden

[a] Presumptive specialization in copulation pattern (see text).

in the initial phases of pairing (Eisenberg, McKay, & Jainudeen, 1971). One further point we should not lose sight of is that the pairing phenomenon between male and female mammals, especially in ones that exhibit a solitary social structure, tends to be somewhat prolonged. It is not simply a brief or casual affair. The male may court actively and remain with the female for a period of 48 hours or more. In the life of a small, short-lived mammal, this is a not inconsiderable interval of time. In such solitary mammals as the tiger (*Panthera tigris*), the male and female may court and mate over a period of several days, and, indeed, considering the low conception rate, such sexual activity may have to be repeated after 45 days when the female again comes into estrus (Kleiman, 1974b).

Parental Care

The Basic Maternal Role

The evolution of mammary glands early on set the pattern for reproduction in the Class Mammalia. In a sense, the female is capable of supplying the newborn young with food which is a by-product of her own metabolism. Although certain bird species (Order Columbiformes) have evolved analogous mechanisms, in the main most reptiles and birds do not provide direct by-products from the body as a nutritive source for the hatched young. When parental care is shown in birds, it generally involves either leading the young to food or provisioning the young with foodstuffs collected by the adults. In altricially hatched birds, such provisioning begins at the time of hatching. In birds, very often the feeding involves the participation of both sexes. This is especially true of those bird species in the Order Passeriformes (see Kendeigh, 1952, for review).

The assumption of early neonatal nutrition as the role of the female mammal must have taken place very early in the mammalian evolutionary history, since this is clearly established in the monotremes, marsupials, and eutherians. This initial separation of roles in initial parental care is so profound in the Mammalia that male involvement in parental care during the early development of the neonate is vastly reduced (see Figure 2).

Consider the monotremes: *Tachyglossus aculeatus*, the echidna, takes approximately 177 days to raise the young from conception to semi-independence. After mating with the male, the female takes approximately 27 days to develop the egg within the oviduct. The egg is laid, apparently with the female curled in such a way as to bring the cloaca opposite her pouch. The egg is deposited in the pouch where it undergoes an incubation period of approximately 10 days duration. During this time the female moves little and remains in her burrow. Upon hatching, the young seeks nourishment from the mammary gland orifices, which open into the pouch. The pouch phase of the young lasts approximately 50 days, whereupon its spines begin to develop. At this time, the young is left in the nest while the mother

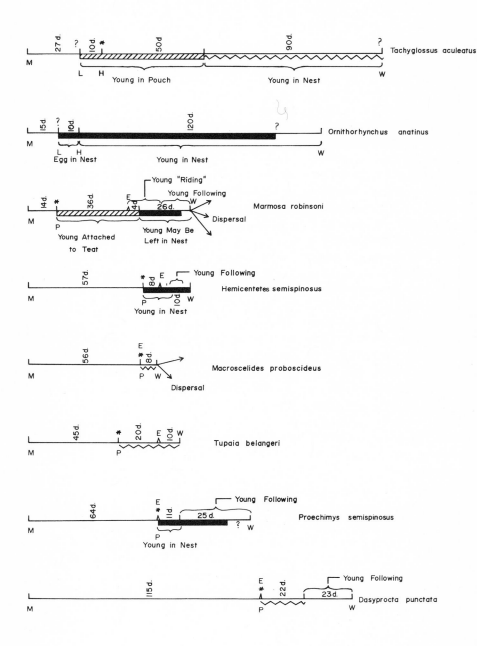

begins to forage more frequently alone, returning to the nest to suckle the young during a period of some 90 days, until the young reaches independence. A single young is generally raised (Griffiths, 1968).

The duckbilled platypus, *Ornithorynchus anatinus*, takes approximately 144 days to rear from one to two young. After mating and fertilization, the female retires to her own burrow system and seals herself in. The egg develops in the oviduct for some 15 days and is laid in a nest. The mother incubates the egg continuously for some 10 days, whereupon it hatches. The young is brooded and nursed in the nest for approximately 120 days, the mother only making excursions to the outside to feed herself (Fleay, 1944).

Turning to the Marsupialia, we may take *Marmosa robinsoni* as a typical developmental cycle. This small animal, seldom exceeding 35 grams adult wright, takes 80 days to raise a litter of around 10 young. After mating, the sexes separate, with the female becoming quite antagonistic toward the male, and the young develop with the aid of a yolk-sac placenta in her uterus for some 14 days. At parturition, the young crawl unaided to the teat area of the mother. In this particular species of opossum, there is no pouch. The young attach to the teat and remain attached for some 30 days, after which the eyes open. Between the 30th and 40th day of age, the young will begin to detach from the teat and at times are transported on the mother's back. The mother forages actively even with pouch young, but, from about 40 days on, the young are left in the nest while the mother forages alone, returning to suckle them. This nest phase for the young lasts approximately 25 days, whereupon they begin to disperse from the nest and initiate an independent life (Eisenberg & Maliniak, 1967; Collins, 1973).

The eutherian tenrec, *H. semispinosus*, takes about 75 days to rear six young to weaning. As with the eutherians, the intrauterine phase is prolonged; 57 days after mating, parturition takes place in a

---

Figure 2. Time sequences of maternal care activities for selected mammals. ⬜⬜⬜⬜ = pouch phase; ∿∿∿ = absentee maternal care phase: ████ = nest phase with maternal brooding; M = mating; H = hatching; L = laying; P = partus; * = beginning of lactation; W = end of weaning period; E = eye opening; d = days. Convergences and parallelisms are indicated when marsupials, monotremes, and eutherians are compared. (See text from Eisenberg, 1975.)

nest. The young are brooded and suckled by the female in her individual nest site for some 14 days, whereupon the young initiate a following response and are guided in their initial foraging by means of signals produced by specialized quills on the female's back, termed *stridulating quills* (Eisenberg & Gould, 1970). The young are weaned, perhaps 22 days postpartum. During the time from 14 to 22 days, the young are acquainted with the area in the vicinity of the female's nest site and are able to orientate themselves to the environment through the initial following response shown with respect to the mother (Eisenberg & Gould, 1970).

The remaining eutherian patterns that I wish to discuss are all characterized by this extended intrauterine phase of development. I would like to point out two variations in the maternal care patterns. One involves the so-called absentee parental care pattern (see Martin, 1968), where the young are born in a special nest site which is visited periodically by the mother for lactation. This system has evolved in a convergent and parallel fashion several times in the eutherian mammals. The elephant shrew, *Macroscelides probosideus*, bears from one to two precocial young in a secluded nest site after a 56-day gestation period. The young are born fully haired with the eyes open. The lactation phase comprises only 11 to 12 days, with the female returning to the nest site to nurse the young and then departing for 24 to 48 hours (Sauer & Sauer, 1972).

The cursorial caviomorph rodent genus, *Dasyprocta*, exhibits a similar life cycle. After a 115-day gestation period, the young, (one to two), are born with the eyes open, generally in a secluded spot, whereupon the female guides the young to a burrow and induces it to enter. The young itself may enlarge the burrow and remains there only to emerge upon the mother's visits for nursing. This hiding–nursing phase may persist for nearly 20 days until the young is strong enough to follow the mother in a manner reminiscent of parental care patterns in many Artiodactylans (Lent, 1974; Smythe, 1970; Kleiman, 1972).

Lest one think that the absentee parental care system is always found in conjunction with precocial young, I should cite the related caviomorph rodent, *Proechimys semispinosus,* which has a litter of one to five young, born with eyes open and fully furred, after a 64-day gestation period. The young are born in a nest within a burrow and are suckled in the burrow for some 38 days prior to weaning.

The young generally do not follow the mother on her nocturnal excursions from the nest until approximately 12 days postpartum. In this case, however, the mother always returns to the same nest site to brood and suckle her litter (Maliniak & Eisenberg, 1971).

Even in highly specialized, recently evolved taxa, the primary female parental role may remain. In the leopard (*Panthera pardus*), the female may carry out all of the primary parental care functions. It would seem that the role of the male, whose home range overlaps that of the female, is in part involved with keeping the area free from other males; the presence of such other males on a long-term basis could considerably reduce the carrying capacity of the area. Thus, the female and her progeny derive indirect benefits through the activities of the male (Muckenhirn & Eisenberg, 1973; Eisenberg & Lockhart, 1972). Furthermore, in species such as the larger cats, the young may remain with the mother for over two years and learn to hunt. The youngsters will simply not be able to survive if they are separated from the mother at too early an age. Thus, a social grouping of two to three individuals is not uncommon in the leopard, but it is a mother family.

The Role of the Male

In all of the preceding examples, we see variations on a common theme, i.e., the initial phases of rearing fall in the main to the female. In spite of this seeming lack of participation, one of the primary roles of the male is to maintain the home range of the female free of males which could compete for resources. As Lockie (1966) has outlined for the Mustelidae, a given female's home range is encompassed by a male's range. In fact, he may even mate with her on subsequent breeding seasons, and, although he does not participate directly in parental care, he certainly reduces competition for prey by keeping the area free from other males.

Brown (1966) comments on similar behavior in small rodents of the genera *Apodemus* and *Clethrionomys*. The home range of an adult male may encompass the home ranges of one or more females. In addition, a dominant adult male has a relatively large home range compared to other subordinate adult males which may partially inhabit the same living space. It appears then that a dominant adult male will tolerate subordinate adult males within his home range but that these males' movements are inhibited such that they do not forage over an area as widely as would be the case if the dominant adult male were not present. Myton (1974) refers to this as a form of "family clustering" and finds the same sort of behavior in the North American rodent genus, *Peromyscus*.

On the other hand, although early nutritive care in the Mammalia invariably falls to the female, the male may be involved in the rearing of the young at a somewhat later stage. Species of the primate family Callithricidae exhibit profound parental care on the part of the male, since the male generally carries the young during their dependent phase or a major portion of it, transferring them to the female for lactation (Epple, 1967). Furthermore, in the family Canidae of the Carnivora, provisioning of the female by the male and subsequent provisioning of the offspring often involve the adult male or the young of the previous year (Kleiman & Eisenberg, 1973).

In the cape hunting dog (*Lycaon pictus*) and the wolf (*Canis lupus*), males actively provision the female, and the integrity of the pack results from two factors: (1) the tendency for only one pair to reproduce within the pack and (2) the slow maturation of the progeny of the founding adults. These half- to three-quarters-grown animals can provide additional foraging ability in provisioning the founding female and her subsequent progeny (Kleiman & Eisenberg, 1973). Furthermore, such a pack of related individuals can hunt more efficiently and, indeed, show significant cooperation during the hunt for small game, resulting in greater efficiency in prey capture (see Schaller & Lowther, 1969; Mech, 1970). These examples are exceptions rather than the rule, so that when the Class Mammalia is surveyed as a whole, primary parental investment appears to fall to the female.

The overall consequence of this differential investment of energy may be related to the theories of Trivers (1972). As he rightly points out, the parental investment of the female mammal is so heavily involved with her own offspring that male replication of genotype is often not involved with the particular defense of a given female, but rather his reproductive success is a function of how many females he can effectively inseminate. Thus, polygynous systems, in their many forms of expression, are more often the rule in the Class Mammalia than is the case in the Class Aves (see Orians, 1969). The consequences of polygynous mating systems have been reviewed many times. It should be pointed out, however, that ritual competition among males for access to several breeding females places a high selective advantage upon males with great size: larger horns, teeth, etc. The competition of males imposes a form of social selection often leading to pronounced sexual dimorphism within species. The problem is ably placed in perspective by Crook (1972).

The Derivation of Complex Social Structures

If the social structures exhibited by a species are viewed as adaptive expressions of behavioral phenotypes (Eisenberg, 1966; Crook & Gartlan, 1966; Crook, 1964), then the reproductive unit as a subset of the species' social system is no less the product of natural selection. Mating systems of the Cervidae and Bovidae have been treated from the standpoint of general adaptations to existing environmental conditions (Eisenberg & Lockhart, 1972, pp. 81 to 90; Jarman, 1974), as well as from the perspective of evolutionary history (Geist, 1971, 1966). Of course, such integrative efforts attempt to determine the "averages" of a species' behavior—they seek to define a behavioral "mode" (Leyhausen, 1965b). It is necessary to keep in mind that the methods of sexual attraction and mating as well as parental care are subject to some interindividual variation, and, at best, a "species typical" form of behavior is an abstraction.

Given the species-typical variations of parental care displayed by females whose primary responsibility is the initial rearing of young, we may note in comparing the various evolutionary lines of mammals that the formation of more complicated social configurations has occurred independently many times. If we leave aside migrating herds and roosting colonies of bats, where mother–infant bonds are usually strong but where, otherwise, individual recognition does not seem to be important, and consider only those social structures where individual recognition is important and where the structure itself exhibits a high degree of cohesion, then we find that, over a wide range of mammalian orders, similar forms of social structure have been evolved in a parallel and/or convergent fashion. What specifically have evolved are the behavioral regulatory mechanisms and modes of communication which permit the control of individuals within a social context without pathological repercussion.

As pointed out in an earlier paper (Eisenberg, 1967), species of mammals which are adapted for a more or less solitary existence may in fact be conditioned to live in groups, but, when this is done, reproductive failure generally results (Eisenberg, 1969). Thus a truly social species is one that has the necessary communicatory mechanisms and modes of interaction which permit tolerance with conspecifics without loss of the potential to reproduce. In short, the social structure so formed is in fact a device for replication of itself, and indeed, in an evolutionary sense, the structure has been favored because it favors the reproduction of the species.

The reproductive unit in the Mammalia becomes more complex in terms of the numbers of interacting individuals, if either one or both of the following steps occur: (1) involvement of the male in parental care and (2) retention of some or all current young with the female through a second rearing phase. If (2) occurs, then often the more mature young assist in some manner in the rearing and socialization of the next age class. In order for such systems to develop, the potential for the expression of interindividual aggression must be controlled. Competition among females must be reduced, as well as competition among males. The latter is apparently accomplished less easily since the very fact that primary parental investment lies with the females means that any given male can increase his fitness by mating with as many females as possible (see Trivers, 1972). Such a mating system, however, inevitably leads to competition among males. Thus it should not be surprising that male–male competition is common and has often led in an evolutionary sense to the formation of ritualized mechanisms (e.g., leks, harems, etc.) for competition among males for access to breeding females (Buechner, 1961; Bartholomew & Hoel, 1953; Koford, 1957; Leuthold, 1966; Geist, 1971).

Why have such different reproducing systems evolved? Clearly, the answer must lie in the overall adaptation of the species for the exploitation of its habitat. As Crook (1970) has pointed out, a multiplicity of influences shapes the form of a species' social organization. The distribution of foodstuffs in space, whether clumped and scattered or broadly distributed, sets the stage for, on the one hand, discrete defense of an area utilized and, on the other hand, virtually no defense of a foraging area. As Jarman (1974) has so elegantly shown, the form of an ungulate social organization is very much a function of its overall adaptation to the environment. Its mode of foraging determines the effective size of the group. The permissibility of area defense by a nonmobile species results in a territorial distribution. Pairing is common among small sedentary browsing ungulates where a male and female defend an area in common and rear their progeny. Larger mobile grazing ungulates show a consistent trend toward the formation of herds.

The form of the antipredator behavior, whether it is built around concealment or individual responses, on the one hand, or, on the other hand, group mobbing effects, profoundly affects the selective advantage for differing forms of social organization (Eisenberg & Lockhart, 1972). Rather than review all the criteria, I refer the reader to Crook's excellent summary (Crook, 1970).

Bearing in mind that the evolution of extended periods of reproductive activity with concomitant longer life and slower maturation rates has often led to the formation of multigenerational social groupings, let us look at some case studies concerning selective advantages that favor evolution of higher social units in the Class Mammalia.

The Matriarchy

One of the more common forms of social organization repeatedly evolved within the Class Mammalia consists of a matriarchy. Essentially, a female and a series of daughters or sisters, age graded, participate mutually in the rearing of their collective progeny. The elephant serves as a typical example.

In a species such as *E. maximus*, the Asiatic elephant, the male participates very little in the initial phases of rearing the young. The most cohesive social structure is the basic matriarchy, a series of related females, raising their calves in common. In the Asiatic elephant, McKay (1973) has shown that the home ranges of these matriarchal herds tend to be distinct and show little overlap. Adult male home ranges overlap considerably with the females'. The males, however, have some input in the parental care system in that they allow younger males, when driven out from the matriarchal herd, to attach themselves to them and thus learn foraging habits, watering places, etc., through association with the adult males. The cow herd itself remains as a very effective antipredator device. Since the young calf would be an easy prey for large predators, the cows through cooperative rearing of the young can provide maximum protection to the young animals during the first four years of their lives (McKay, 1973).

An incipient form of such a social structure may be shown by the tenrec, *H. semispinosus*. The tenrec configuration generally involves the communal use by several females of a burrow system, but they forage independently. The elephant system actually involves a functional subdivision of the herd into groups of females with young of a similar age class. The female elephants in part act as a defensive unit for the young, and the elephant social structure is necessary for the rearing of young. The tenrec female *Hemicentetes* may in fact rear her young alone, but it is highly dubious that an elephant female would be capable of the same act. It would appear that different selective pressures have promoted similar social configurations, when these two distantly related species are compared. The elephant matriarchy shows cohesion as it moves about and feeds together. In the tenrec, however, collective nesting appears to occur only under conditions where the foraging area for evening feeding is sufficiently large to support a great many tenrecs, while only an extensive burrow system is used in common by several females. If the burrow of the tenrecs is inadvertently broken open by a predator, individual defensive reactions by several females may prove more of an antipredator deterrent than those by a single adult female. Whether this is of any selective advantage in promoting sociality in *H. semispinosus* is at this point speculative. It would appear, however, that the capacity to use restricted burrow sites for communal rearing has some adaptive advantage (Eisenberg & Gould, 1970).

A Matriarchy with Male Attachment

In a similar fashion, lion (*Panthera leo*) prides have evolved essentially because of the mutual advantage accruing to females who hunt communally, since their hunting success is higher when several females hunt together than when one hunts alone (Schaller, 1972). Although each female individually rears her cubs until they are of an age to follow, during the later phases of rearing females hunt together and no doubt increase the chances of cub survivorship by communal feeding at the kill. It would appear that the male lion's primary role in lion society is to keep the area clear of strange males, thus maintaining a higher carrying capacity for the reproductive females and reducing the amount of interference to the females and cubs from strange males.

Packs as Extended Families

In the case of group-living canids, such as *Canis lupus, Lycaon pictus*, and *C. aureus*, we have an extended family based on a single reproductive pair and their progeny of the previous breeding season. In the case of the wolf pack, delayed sexual maturation on the part of young males may maintain them in the group for two or three years. In the case of the jackal, with the onset of sexual maturity the young males and often young females disperse immediately. Thus, the jackal maintains what appears to be a pair configuration or pair with young, while the wolf may show a larger grouping. Nevertheless, the basic structure is quite similar. In these group-living canids, participation in hunting permits them to obtain larger prey than would necessarily be the case if they hunted alone. The male actively provisions the female and her cubs, and provisioning is a group effort, with parents and subadults provisioning younger animals. In all cases, however, there is generally only one reproducing pair of animals, the founding father and mother, within the so-called pack (see Kleiman & Eisenberg, 1973).

The Influence of Phylogeny

Some mammalian taxa exhibit rather uniform trends in the expression of their social organization, which may in fact be related to a single overriding environmental parameter, interplaying with the unique phylogenetic background of the family in question. If we compare two families of desert rodents, the Dipodidae (jerboas), on the one hand, and the Heteromyidae (kangaroo rats and pocket mice), on the other, this point can be made clear. The kangaroo rats evolved in the deserts of North America convergent to the evolution of the jerboas in Central Asia and North Africa. The dipodids have apparently had the longer evolutionary history and in many respects are more specialized. The two families show in common the following attributes: bipedal locomotion, highly developed techniques for gathering seeds from sand, similar forms of burrow construction; but they differ profoundly in that most of the dipodids accumulate large fat reservoirs and hibernate during the winter, whereas the kangaroo rats do not accumulate fat reservoirs, do not hibernate, and instead cache vast quantities of seeds in individually defended burrow systems. The seed-caching habit and the high development of potential to express aggressive behavior results in individual occupancy of burrows for all adult kangaroo rats (Eisenberg, 1963).

On the other hand, in the family Dipodidae, greater degrees of social tolerance may be shown. Apparently selection has not favored mechanisms to ensure complete burrow defense, probably as a response to relaxed selective pressure for defense of a cache. Thus, a family such as the North American Heteromyidae may show an enduring trend toward a solitary dispersed social system, whereas the convergently evolved Dipodidae show more variations on the theme, presumably because their overwintering system is not based upon a seed-caching pattern (Eisenberg, 1967).

To summarize then, the social structures manifested by a given species are the product of selective pressures, which vary from one species to the other as a function of the particular ecological niche they exploit. Given the basic mammalian reproductive unit, in the course of adaptation the mammals have evolved alternative expressions based upon either (1) extending the group by retaining females within it, creating a matriarchy, or (2) incorporating the male in some form of parental care, thus constituting a nuclear family, or (3) a combination of both. The step beyond this is to incorporate several males and several females into an extended rearing group, and this has been accomplished within primates, cetaceans, and some ungulates (Eisenberg, 1966). Even so, there is almost invariably a hierarchy among the males and females with differential reproductive success depending upon one's position in the hierarchy. Generally in such social configurations more dominant males have a greater reproductive success. To some extent the trend is masked in the female hierarchy.

The Multimale System

In spite of the apparent differences in such multimale and multifemale social groupings, they are in fact variations on the same theme of extended nuclear families, and almost invariably the affiliation mechanisms among the members result from a developmental history within the same kin group. I think as a rule of the thumb it is fair to say that the female band or female component of such social configurations shows the greatest stability and that young males tend to move between groups, thus introducing outbreeding and reducing the potential for extensive inbreeding.

The manner in which young males can transfer from one social group to the next is variable. In primates, such as the langurs of the genus *Presbytis*, the takeover is often violent with loss of young and serious wounding through male fighting (Rudran, 1973). [A similar phenomenon has been noted for the lion (*Panthera leo*) (B. Bertram, personal communication).] In species of primates such as the Ceylon macaque, *Macaca sinica*, the takeover may be gradual and may in-

Table II. Glossary of Common Names

Antechinus stuarti, Stuart's marsupial mouse	*Marmota olympus*, Olympic marmot
Apodemus, Old World wood mouse	*Megaleia rufa*, red kangaroo
Blarina brevicauda, short-tailed shrew	*Mesocricetus auratus*, golden hamster
Canis aureus, golden jackal	*Microgale dobsoni*, Dobson's long-tailed tenrec
Canis lupus, wolf	*Microgale talazaci*, long-tailed tenrec
Clethrionomys, red back vole	*Mus musculus*, house mouse
Cynomys ludovicianus, black-tailed prairie dog	*Ornithoryhnchus anatinus*, platypus
Dasyprocta, agouti	*Oryctolagus*, rabbit
Didelphis marsupialis, Virginia opossum	*Panthera leo*, lion
Echinops telfairi, lesser hedgehog tenrec	*Panthera pardus*, leopard
Elephas maximus, Asiatic elephant	*Panthera tigris*, tiger
Hemicentetes semispinosus, streaked tenrec	*Peromyscus*, deer mouse
Lycaon pictus, African hunting dog	*Presbytis*, langur monkey
Macaca sinica, toque macaque	*Proechimys semispinosus*, spiny rat
Macropus gigantea, grey kangaroo	*Setifer setosus*, giant hedgehog tenrec
Macroscelides proboscideus, elephant shrew	*Sminthopsis crassicaudata*, fat-tailed marsupial mouse
Marmosa mitis, masked mouse opossum	*Suncus murinus*, musk shrew
Marmosa robinsoni, mouse opossum	*Tachyglossus aculeatus*, echidna
Marmota, woodchucks and marmots	*Tamiasciurus*, North American red squirrels
Marmota flaviventris, yellow-bellied marmot	*Tenrec*, common tenrec
Marmota monax, woodchuck	

volve mutual support among males seeking to attach themselves to a new troop with reduced antagonism (Dittus, 1974).

This brings us then to the question of the genesis of multimale units, that is, social structures where several males are combined with several females into a cohesive foraging unit or, if it subdivides, then a unit where great mutual tolerance is shown when the subdivisions come together. Although multimale social groupings are found in several mammalian orders, the question is most controversial in the order Primates. In an earlier publication (Eisenberg *et al.*, 1972), it was suggested that, although this form of social structure appears to have evolved convergently several times within the primates, many of such multimale societies that have been described are more apparent than real. It was suggested that the multimale condition in some species may in fact be viewed as an age-graded male system with a dominant founding male in reality doing most of the breeding. His nearly grown sons by their presence may contribute to successful antipredator defense or competitive defense against conspecific troops but, in fact, be dominated and graded according to their age. Ultimately, emigration by nearly adult or adult males would result in the formation of new troops either by stealing females or by taking over established troops while displacing the resident adult and his cohorts. Some credence for his view is to be derived from the research on monkeys of the genus *Presbytis*. As outlined by Rudran (1973), the structure of langur troops may in fact reflect differential responses to density, and the male takeover so dramatically demonstrated in these species may occur with a frequency that parallels the density of the troops themselves. While I do not mean to imply that the genesis of *Presbytis* groupings is typical for all primates which exhibit multimale or age-graded male systems, I do want to suggest that trends toward polygyny and exclusive access to females during the time of ovulation by a single male may be more widespread than is supposed.

Future Problems

Although we have talked a great deal about discrete social systems as mechanisms for self-replication, it should be pointed out that in natural populations these reproductive systems are functionally organized within an ecosystem to constitute a deme. The deme is a temporarily reproductively isolated collection of individuals which may produce young of at least two behavioral phenotypes, young which will be integrated as replacements in the founding area and young that emigrate to found new groups in new areas. This, in fact, is the real unit of reproduction as so ably pointed out by Anderson (1970). While I do not want to make excusions into population genetics at this point, I only wish to indicate that the discrete social structures which we have been talking about are components of a larger integrated system for self-replication, which we have only begun to think about, let alone describe and classify.

References

Anderson, P. K. Ecological structure and gene flow in small mammals. *Symposium of the Zoological Society of London, 26*, 299–352. New York: Academic Press, 1970.

Armitage, K. B. Social behaviour of a colony of the yellow-bellied marmot (*Marmota flaviventris*). *Animal Behaviour*, 1962, *10*, 319–331.

Barash, D. P. The social biology of the Olympic marmot. *Animal Behaviour Monographs*, 1973, *6*(3), 171–245.

Barnes, R. D., & Barthold, S. W. Reproduction and breeding behavior in an experimental colony of *Marmosa mitis. Journal of Reproduction and Fertility, Supplement*, 1969, *6*, 477–482.

Bartholomew, G., & Hoel, P. G. Reproductive behavior of the Alaska fur seal, *Callorhinus ursinus. Journal of Mammalogy*, 1953, *34*, 417–436.

Bronson, E. H. Agonistic behavior in woodchucks. *Animal Behaviour*, 1964, *12*, 470–478.

Brown, J. L. The evolution of diversity in avian territorial systems. *Wilson Bulletin* 1964, *6*, 160–169.

Brown, L. E. Home range and movement of small mammals. pp. 111–142. In P. A. Jewell & C. Loizos (Eds.), *Play, exploration* and *territory in mammals*. New York: Academic Press, 1966.

Buechner, H. K. Territorial behavior in Uganda kob. *Science*, 1961, *133*, 698–699.

Burt, W. Territoriality and home range concepts as applied to mammals. *Journal of Mammalogy*, 1943, *24*, 346–352.

Collins, L. R. Monotremes and marsupials: A reference for zoological institutions. Smithsonian Press, 1973.

Crook, J. H. The evolution of social organization and visual communication in weaver birds. *Behaviour Supplement*, 1964, *10*, 1–178.

Crook, J. H. The socio-ecology of primates, pp. 103–166. In J. H. Crook (Ed.), *Social behavior in birds and mammals*. New York: Academic Press, 1970.

Crook, J. H. Sexual selection, dimorphism, and social organization in the primates, pp. 231–281. In B. Campbell (Ed.), *Sexual selection and the descent of man.* Chicago: Aldine, 1972.

Crook, J. H., Gartlan, S. Evolution of primate societies. *Nature*, 1966, *210*, 1200–1203.

Deegener, P. *Die Formen der Vergesellschaftung im Tierreiche.* Leipzig: von Veit, 1918.

Devor, M., & Murphy, M. The effect of peripheral olfactory blockage on the social behavior of the male golden hamster. *Behavioral Biology*, 1973, *9*, 31–42.

Dittus, W. The ecology and social behavior of the toque macaque (*Macaca sinica*). Unpublished Ph.D. thesis, University of Maryland, 1974.

Downhower, J., & Armitage, K. B. The yellow-bellied marmot and the evolution of polygamy. *American Naturalist*, 1971, *105*, 355–370.

Dryden, G. Reproduction in *Suncus murinus. Journal of Reproduction and Fertility, Supplement*, 1969, *6*, 377–396.

Eisenberg, J. F. Behavior of heteromyid rodents. *University of California Berkeley, Publications in Zoology*, 1963, *69*, 1–100 + 13 plates.

Eisenberg, J. F. The social organizations of mammals. *Handbuch der Zoologie*, 1966, *8* (10/7), Lieferung 39.

Eisenberg, J. F. Comparative studies on the behavior of rodents with special emphasis on the evolution of social behavior, Part I. *Proceedings of the U.S. National Museum*, 1967, *122* (3597), 1–55.

Eisenberg, J. F. Social organization and emotional behavior. In E. Tobach (Ed.), *Experimental approaches to the study of emotional behavior. Annals of the New York Academy of Sciences*, 1969, *159*(3), 752–760.

Eisenberg, J. F. Phylogeny, behavior, and ecology in the Mammalia, pp. 47–68. In P. Luckett & F. Szalay (Eds.), *Phylogeny of the primates: An interdisciplinary approach.* New York: Plenum Press, 1975.

Eisenberg, J. F., & Gould, E. The tenrecs: A study in mammalian behavior and evolution. *Smithsonian Contributions to Zoology*, 1970, *27*, 1–137.

Eisenberg, J. F., & Kleiman, D. G. Olfactory communication in mammals. *Annual Revue of Ecology & Systematics*, 1972, *3*, 1–32.

Eisenberg, E. F., & Lockhart, M. An ecological reconnaissance of Wilpattu National Park, Ceylon. *Smithsonian Contributions to Zoology,* 1972, (101).

Eisenberg, J. F., & Maliniak, E. The breeding of *Marmosa* in captivity. *International Zoo Yearbook*, 1967, *7*, 78–79.

Eisenberg, J. F., McKay, G., & Jainudeen, M. R. Reproductive behavior of the Asiatic elephant (*Elephas maximus* L.). *Behaviour*, 1971 *38*, 193–225.

Eisenberg, J. F., Muckenhirn, N., & Rudran, R. The relationship between ecology and social structure in primates. *Science*, 1972, *176*, 863–874.

Epple, G. Vergleichende untersuchungen über Sexual- und Sozialverhalten der Krallenaffen (Hapalidae). *Folia Primatologica*, 1967, *7*, 37–65.

Ewer, R. R. *Ethology of mammals.* London: Logos Press, 1968.

Fisler, G. F. Mammalian organizational systems. *Los Angeles County Museum Contributions in Science*, 1969, (167).

Fleay, D. *We breed the platypus.* Melbourne: Robertson & Mullens, 1944.

Geist, V. Ethological observations on some North American cervids. *Zoologische Beiträge* (n.s.), 1966, *12*, 219–250.

Geist, V. *The mountain sheep.* University of Chicago Press, 1971.

Gould, E. Communication in the three genera of shrews (Soricidae): *Suncus, Blarina* and *Cryptotis. Communication Behavior Biology A.*, 1969, *3*, 11–31.

Griffiths, M. *Echidnas.* London: Pergamon, 1968.

Hediger, H. Säugetiere Soziologie. In *Structure et physiologie des societies animales.* Paris: C.N.R.S., 1952.

Jarman, P. J. The social organization of antelope in relation to their ecology. *Behaviour*, 1974, *48*, 215–267.

Jerison, H. J. *Evolution of the brain and intelligence.* New York: Academic Press, 1973.

Kendeigh, S. C. Parental care and its evolution in birds. *Illinois Biological Monographs*, 1952, *22* (1–3).

King, J. A. Social behavior, social organization, and population dynamics in a black-tailed prairiedog town in the Black Hills of South Dakota. *Contributions from the Laboratory of Vertebrate Biology of the University of Michigan*, 1955, 67.

Kleiman, D. G. Maternal behaviour of the green acouchi, *Myoprocta pratti*, a South American caviomorph rodent. *Behavior*, 1972, *43*, 48–84.

Kleiman, D. G. Patterns of behavior in hystricomorph rodents. In I. W. Rowlands & B. J. Weir (Eds.), *The biology of hystricomorph rodents. Symposium of the Zoology Society of London*, 1974a, *34*, 171–209.

Kleiman, D. G. The estrous cycle in the tiger, pp. 60–75. In R. H. Eaton (Ed.), *The world's cats.* Vol 2. Seattle: Feline Research Group, 1974b.

Kleiman, D. G., & Eisenberg, J. F. Comparisons of canid and felid social systems from an evolutionary perspective. *Animal Behaviour*, 1973, *21*, 637–659.

Klingle, H. Das Verhalten der Pferde. *Handbuch der Zoologie*, 1972, *8, 10*(24), 1–66.

Koford, C. B. The vicuña and the puna. *Ecological Monographs*, 1957, *27*, 153–219.

Krumbiegel, I. *Biologie der Säugetiere.* (2 vols.) Krefeld: Agis Verlag, 1953–54.

Lent, P. Mother–infant relationships in ungulates, pp. 14–55. In V. Geist & F. Walther (Eds.), *The behavior of ungulates and its relation to management*, Vol. 1. I.U.C.N. Publ. 24, Morges, 1974.

Leuthold, W. Variations in territorial behavior of Uganda kob *Adenota kob thomasi* (Neumann 1896). *Behavior*, 1966, *27*, 214–257.

Leyhausen, P. The communal organization of solitary mammals. *Symposium of the Zoological Society of London*, 1965a, *14*, 249–263.

Leyhausen, P. Über die Funktion der Relativen Stimmungshierarchie. *Zeitschrift für Tierpsychologie*, 1965b, *22*, 412–494.

Lockie, J. D. Territory in small carnivores. *Symposium of the Zoological Society of London*, 1966, *18*, 143–165.

Maliniak, E., & Eisenberg, J. F. The breeding of *Proechimys semispinosus* in captivity. *International Zoo Yearbook*, 1971, *11*, 93–98.

Martin, R. D. Reproduction and ontogeny in tree shrews (*Tupaia belangeri*) with reference to their general behavior and taxonomic relationships. *Zeitschrift für Tierpsychologie*, 1968, *25*(4), 409–495.

McBride, G. A general theory of social organization and behavior. *Faculty Veterinary Science*, 1964, *1*(2), 75–100. St. Lucia, Brisbane: University of Queensland Press.

McKay, G. M. Behavior and ecology of the elephant (*Elephas maximus*) in southeastern Ceylon. *Smithsonian Contribution, in Zoology*, 1973, *125*, 1–113.

Mech, D. *The wolf*. NewYork: Natural History Press, 1970.

Muchkenhirn, N. A., & Eisenberg, J. F. Home ranges and predation in the Ceylon leopard, pp. 142–175. In R. L. Eaton (Ed.), *The world's cats*, Vol. 1., *Ecology and conservation*. Winston, Oregon: World Wildlife Safari, and Athens, Georgia: ISCES, 1973.

Murphy, M. Effects of female hamster vaginal discharge on the behavior of male hamsters. *Behavioral Biology*, 1973, *9*, 367–375.

Myton, B. Utilization of space by *Peromyscus leucopus* and other small mammals. *Ecology*, 1974, *55*, 277–290.

Orians, G. H. On the evolution of mating systems in birds and mammals. *American Naturalist*, 1969, *103*, 589–603.

Pearson, O. P. Reproduction in the shrew (*Blarina brevicauda* Say). *American Journal of Anatomy*, 1944, *75*, 39–93.

Portman, A. Über die Evolution der Tragzeit bei Saugetieren. *Revue Suisse de Zoologie*, 1965, *72*, 658–666.

Reynolds, H. C. Studies on reproduction in the opossum (*Didelphis v. virginiana*). *University of California Publication in Zoology*, 1952, *52*(3), 223–284.

Richmond, M., & Conaway, C. H. Induced ovulation and oestrous in *Microtus ochrogaster*. *Journal of Reproduction and Fertility, Supplement*, 1969, *6*, 357–376.

Rudran, R. Adult male replacement in one-male troops of purplefaced langurs (*Presbytis senex senex*) and its effect on population structure. *Folia Primatologica*, 1973, *19*, 166–192.

Sauer, E. G., & Sauer, E. M. Zur biologic der kurzohrigen Elefantenspitzmaus in der Namib. *Zeitschrift des Kölner Zoo*, 1972, *15*(4), 119–139.

Schaller, G. *The serengeti lion*. University of Chicago Press, 1972.

Schaller, G., & Lowther, G. R. The relevance of carnivore behavior to the study of early hominids. *Southwest Journal of Anthropology*, 1969, *25*(4), 302–341.

Seidensticker, J. C., Hornocker, M. G., Wiles, W. V., and Messick, J. P. Mountain lion social organization in the Idaho primitive area. *Wildlife Monograph* (35), Wildlife Society, 1973.

Sharman, G. B. Marsupials and the evolution of viviparity, pp. 1–28. In J. D. Carthy & C. L. Duddington (Eds.), *Viewpoints in biology*. 4. London: Butterworths, 1965.

Sharman, G. B., Calaby, J. H. and Poole, W. E. Patterns of reproduction in female diprotodont marsupials. *Symposium of the Zoological Society of London*, 1966, *15*, 205–232.

Smith, C. The adaptive nature of social organization in the genus of tree squirrels, *Tamiasciurus*. *Ecological Monographs*, 1968, *38*, 31–63.

Smythe, N. Ecology and behavior of the agouti (*Dasyprocta punctata*) and related species on Barro Colorado Island, Panama. Ph.D. thesis, University of Maryland, 1970.

Stodart, E. Management and behavior of breeding groups of the marsupial *Permeles nasuta* Geoffroyi in captivity. *Australian Journal of Zoology*, 1966, *14*, 611–623.

Trivers, R. L. Parental investment and sexual selection, pp. 136–179. In B. Campbell (ed.), *Sexual selection and the descent of man: 1871–1971*. Chicago: Aldine, 1972.

Wilson, E. O. Group selection and its significance for ecology. *Bioscience*, 1973, *23*, 631–638.

4

Reproductive Isolation, Behavioral Genetics, and Functions of Sexual Behavior in Rodents

Thomas E. McGill

This chapter consists of several interrelated parts. First, Mayr's classification scheme for reproductive isolating mechanisms is reviewed. For several of these mechanisms, behavior is known to play an important role, while for others the importance of behavioral differences has not been demonstrated. Much of the remainder of the chapter is concerned with the possibility that behavior, specifically the copulatory act itself, might function in these latter kinds of reproductive isolation. Since the examples offered deal with rodent sexual behavior, the second section of the chapter examines some species differences in sexual responses within that order.

The third section reviews behavior–genetic questions related to the evolution and function of reproductive behavior and some representative studies involving both laboratory strains and naturally occurring populations of rodents. The possibility that quantitative differences within species-typical patterns of sexual response might represent incipient reproductive barriers is considered.

Next, differences in the manner by which male Norway rats and

Thomas E. McGill · Department of Psychology, Williams College, Williamstown, Massachusetts 01267.

male house mice induce the luteal phase of the estrous cycle in females are described. The final section of the paper relates these findings to a laboratory discovery called the *enforced-interval effect*. Speculations and hypotheses concerning the adaptive significance of these phenomena in the natural situation are included.

Introduction

In a paper dedicated to Konrad Lorenz on the occasion of his 60th birthday, Niko Tinbergen (1963) outlined four interrelated problem areas for those engaged in the study of animal behavior. These four problems are shown in the left side of Table I.

Questions regarding causation of behavior involve both the internal physiological condition of the organism and external stimuli, animate and inanimate.

The development or ontogeny of behavior constitutes the next problem area. It is here that the sterile nature-versus-nurture controversy continually erupts.

The third problem is that of the evolution of behavior. The concern is with the course evolution has taken and with the dynamics by which it has achieved its ends. The first aspect is pursued through comparative studies of closely related species. Information concerning the dynamics of the evolutionary process is obtained from studies of subspecies and strains, of the effects of point mutations on behavior, and of crossbreeding and selection experiments.

The fourth area, one which Tinbergen has studied extensively, is that of the function or survival value of behavioral patterns. It is to questions regarding the function of behavioral patterns that this

Table I. Tinbergen's Four Problem Areas and Three Behavior–Genetic Questions

Animal Behavior	
Causation	Genotype differences (strains and sub-
Development	species)?
Evolution	Heritability (variances)?
Function	Inheritance mode (means)?

chapter primarily is directed. Tinbergen notes that in the post-Darwin era there was a reaction against uncritical speculations concerning the survival value of anatomical structures as well as behavioral patterns. This reaction degenerated into what Tinbergen called "an attitude of intolerance; even wondering about survival value was considered unscientific." He credits a revival of interest in the study of the functions of behavior to Lorenz's early work. However, it was for Tinbergen himself to demonstrate that meaningful experiments involving survival value could be performed in the field. Perhaps the best example is the elegant work on eggshell removal in black-headed gulls (Tinbergen, Broekhuysen, Feekes, Houghton, Kruuk, & Szulc, 1962; Tinbergen, Kruuk, & Paillette, 1962).

In the study of function, emphasis is on the effects of behavior on future events rather than on the past events that caused the behavior itself. Behavior can influence future events by changing the probability that these events will occur. Behavior such as eggshell removal in gulls can decrease the probability of predation, while behavior such as territory defense can increase the probability of securing a mate. Behavior can simultaneously increase the probability of intraspecific matings and decrease the probability of interspecific matings.

I should like to make one further point regarding Tinbergen's four questions. Few of us have the talent to contribute importantly to all of these areas, but Daniel S. Lehrman was an exception. The major impact of his experimental work with ring doves was in the areas of the causation of behavior and of its functions. However, perhaps equally important were Lehrman's theoretical statements on the problems of the evolution of behavior and, to a greater extent, those of behavioral ontogeny. Indeed, one of his last papers (1970)—a chapter that can be read and reread for both fun and profit—was entitled "Semantic and Conceptual Issues in the Nature–Nurture Problem."

Reproductive Isolating Mechanisms

In the April 1973 issue of *The Auk*, the American Ornithologists' Union reported the demotion of 13 bird species to subspecies rank (Committee, 1973). For example, the myrtle warbler, the blue

Table II. Classification of Isolating Mechanisms[a]

1. Mechanisms that prevent interspecific crosses (premating mechanisms)
    a. Potential mates do not meet (seasonal and habitat isolation)
    b. Potential mates meet but do not mate (ethological isolation)
    c. Copulation attempted but no transfer of sperm takes place (mechanical isolation)

2. Mechanisms that reduce full success of interspecific crosses (postmating mechanisms)
    a. Sperm transfer takes place but egg is not fertilized (gametic mortality)
    b. Egg is fertilized but zygote dies (zygote mortality)
    c. Zygote produces an F_1 hybrid of reduced viability (hybrid inviability)
    d. F_1 hybrid zygote is fully viable but partially or completely sterile or produces deficient F_2 (hybrid sterility)

[a] From Mayr, 1976, p. 57.

goose, the Baltimore oriole, and the slate-colored junco no longer exist as distinct species. The number of bird species was reduced because the animals were quite successfully and happily interbreeding with similar species in areas of overlap; the offspring of these crosses were both viable and fertile.*

These birds were "busted" because they violated the definition of a *species*: "Species are groups of interbreeding natural populations that are reproductively isolated from other such groups" (Mayr, 1970, p. 12). Devices which maintain species integrity are called *reproductive isolating mechanisms*.

Mayr (1963, 1970) has provided a useful sevenfold classification for reproductive isolating mechanisms, as shown in Table II. We will briefly review this scheme, noting the points at which behavior is known or suspected to be an important variable. The first three mechanisms operate to prevent mating and thus are referred to as premating mechanisms.

Seasonal and Habitat Isolation

Obviously, behavioral differences between species are extremely important in maintaining seasonal or habitat isolation. Of the behav-

---

* As an enthusiastic bird watcher, Dan Lehrman would have been very much interested in this report and doubtless amused by the cries of anguish that are arising from certain birders because their life-lists have been suddenly reduced (Arbib, 1973).

ioral mechanisms involved in this and in other types of reproductive isolation, imprinting and other forms of learning are very important. It is perhaps appropriate to note that the last decade has seen a significant increase in research on the importance of learning in the normal life of animals. Not all psychologists have stopped using animals as substitutes for humans in a search for laws of learning, but more and more investigators are considering the role that learning may play in such items as food, habitat, and mate selection. In addition, there is specific recognition that an animal may be genetically prepared to learn some things and quite unprepared to learn others (Hinde & Stevenson-Hinde, 1973; Seligman & Hager, 1972). We can expect progress in our knowledge of the ways in which animals elect to live in certain neighborhoods, eat certain foods, and fall in love with members of their own species.

Ethological Isolation

These barriers are exclusively behavioral and, according to Mayr, "ethological barriers to random mating constitute the largest and most important class of isolating mechanisms in animals" (1970, p. 58). A very great amount of excellent research has been devoted to the study of behavioral barriers to interspecific matings, such as, for example, the work on *Drosophila* courtship patterns by Bastock (1967), Ehrman (1961, 1962, 1965), Manning (1959, 1965, 1971), Spieth (1952), Spiess (1970), and Parsons (1973); Blair's investigations of the mating calls of amphibians (1964); Hinde's review of the mechanisms of speciation in birds (1959); and the work of Alexander (1960, 1962) and others on the songs of crickets. Ethological barriers most frequently depend upon distance receptors. Thus the sounds that an animal makes, the visual stimulation offered by its displays, or the chemical stimuli that it produces can function to prevent copulation between members of different species.

Mechanical Isolation

This isolating mechanism depends upon the fact that the genitals of different sexes of closely related species may not "fit." According to Mayr (1970), mechanical isolation, while originally considered to be of great importance, probably occurs in only a few instances. For certain species, it is possible that behavioral differences might play a part in mechanical isolation even in the absence of major anatomical differences, since the motor patterns involved in gaining intromission and in the act of intromission itself vary greatly.

The next four classes are postmating mechanisms.

Gametic Mortality. In some species antigenic reactions in the genital tract of the female take place; in others the sperm is incapable of penetrating the egg of the alien species.

Zygotic Mortality. Fertilization occurs, but the zygote does not survive. As far as I have been able to determine, there are no examples of behavioral differences producing gametic or zygotic mortality. I will suggest mechanisms by which such reproductive isolation might occur. Whether these mechanisms have ever operated under either laboratory or field conditions is yet to be determined.

Hybrid Inviability. Many hybrids leave no offspring even though they seem to have normal eggs and sperm. It has been suggested that ecological and behavioral differences reduce their chances of successfully reproducing (Mayr, 1970).

Hybrid Sterility. Hybrids are sometimes sterile, and even when this is not the case, cytological, behavioral, or genetic difficulties may appear in the F_2 and backcross groups.

In summary, behavioral differences are of prime importance in seasonal and habitat isolation and in ethological isolation. They are suspected to play a role in hybrid inviability and in hybrid sterility, and they could conceivably operate in mechanical isolation. As mentioned, mechanisms by which behavioral differences could contribute to reproductive isolation through gametic and/or zygotic mortality will be suggested.

One final point should be made. Mayr has emphasized the fact that, although one particular kind of reproductive isolating mechanism may be dominant, it is probable that in most closely related species several mechanisms are capable of operating. He likens the mechanisms to a series of hurdles where, if one is crossed, the next must be overcome.

Species Differences in Rodent Sexual Behavior

In many birds, insects, and fish, elaborate courtship patterns are frequently found. The act of copulation itself is usually relatively simple. To a microsmatic human observer, most rodent species appear to engage in relatively little courtship behavior, while the copulatory act may be a lengthy and complicated process.

Dewsbury (1972) has recently reviewed patterns of copulatory behavior in a variety of male mammals. He proposed a classification scheme based on four questions that can be answered *yes* or *no*. These questions are:

1. Do the male and female become firmly locked together by a strong mechanical connection during copulation?
2. Is there more than one pelvic thrust during intromission?
3. Do multiple intromissions without sperm transfer occur prior to ejaculation?
4. Do multiple ejaculations occur?

Table III shows the results for certain rodents. All of the data in Table III are from Dewsbury (1972), with the exception of those from the two *Clethrionomys* species, which I have added from unpublished observations.

This abbreviated table does not show any rodent species which exhibits locks, but Dewsbury indicates that at least three species do so.

There are several interesting species differences in the pattern of intromission. One which is discussed in greater detail below is that between Norway rats and house mice. The intromission pattern of rats consists of a single pelvic thrust. House mice, on the other hand, execute an average of about 30 thrusts during each intromission (McGill, 1970a), although 10 times that number have occasionally been observed. Note that one of the *Peromyscus* species exhibits multiple thrusts, while all the other species represented for that genus execute a single thrust during intromission. *Microtus* voles have multiple thrusts while *Clethrionomys* voles do not.

Most species exhibit multiple intromissions prior to ejaculation, although there are exceptions such as the California mouse and the meadow vole. The answer to the question regarding multiple ejaculations is more a matter of definition than are the answers to the other three questions. Dewsbury suggests that if copulation resumes within an hour of the initial ejaculation, the male should be considered a *multiple ejaculator*. Rats, for example, exhibit an average of about seven ejaculations prior to *sexual exhaustion*, defined as 30 minutes

Table III. Patterns of Copulatory Behavior in Selected Rodents[a]

Family and species	Common name	Lock?	Thrusting?	Multiple intro- missions?	Multiple ejacu- lations?
Family *Muriade*					
Rattus norvegicus	Norway rat	no	no	yes	yes
Mus musculus	House mouse	no	yes	yes	yes (no)
Family *Cricetidae*					
Mesocricetus auratus	Hamster	no	no	yes	yes
Meriones unguiculatus	Gerbil	no	no	yes	yes
Peromyscus maniculatus	Deer mouse	no	no	yes	yes
P. polionotus	Old-field mouse	no	no	yes	yes
P. gossypinus	Cotton mouse	no	no	yes	yes
P. californicus	California mouse	no	yes	no	yes
P. nasutus	Rock mouse	no	no	yes?	yes?
P. truei	Piñon mouse	no	no	yes?	yes?
Microtus montanus	Mountain vole	no	yes	yes	yes
M. pennsylvanicus	Meadow vole	no	yes	no	yes
M. ochrogaster	Prairie vole	no	yes	yes	yes
Clethrionomys gapperi	Red-backed vole	no	no	yes	yes
C. glareolus	Bank vole	no	no	yes	yes
Family *Caviidae*					
Cavia porcellus	Guinea pig	no	yes	no	no?
Family *Chinchillidae*					
Chinchilla lanigera	Chinchilla	no	yes	no	yes?

[a] Adapted from Dewsbury, 1972. Data for *Clethrionomys gapperi* and *C. glareolus* have been added.

without mounting the female (Beach & Jordon, 1956). Dewsbury designated house mice as multiple ejaculators. However, some of our observations indicate that unless a fresh female is supplied most males do not achieve a second ejaculation within 1 hour. When presented with a female whose vagina is not occluded by the large copulatory plug deposited by male mice during male ejaculation, males of certain genotypes are capable of 2 ejaculations within 1 hour. I have placed the word *no* in parentheses in Table III to indicate this semantic difficulty.

Behavior–Genetic Questions and Experiments

The classification scheme in Table III outlines the species-specific pattern of copulation for certain rodents. Behavior–genetic research is usually concerned with quantitative differences within a species-specific pattern. There are three questions frequently asked in behavior–genetic research which, when properly applied, can assist in the study of all of Tinbergen's problem areas. However, they have particular importance in the study of the function and evolution of behavior. They are indicated on the right side of Table I.

1. Do genotypic differences in behavior exist? For mammals, this question has most frequently been answered by examining different domestic strains, inbred or otherwise. In almost all cases the answer is a resounding *yes*. As I have reported elsewhere (McGill, 1969), most behavior geneticists agree that, if enough strains are studied, significant differences will be found for any quantifiable element of behavior in which one might be interested. Since domestic strains were derived from wild populations, these differences provide strong evidence that genotypically determined behavioral differences exist between individuals or subpopulations of wild animals.

2. The second question concerns heritabilities. Heritabilities are determined by examination of components of variation within populations. The term has two meanings that are sometimes confused. In addition, the concept has frequently been misinterpreted, or indeed reified, in recent controversies regarding human intelligence. Therefore, it is important to understand its limitations as well as its usefulness.

As mentioned, there are two ways of defining heritability. Heritability in the broad sense, or what Falconer (1960, p. 146) perhaps more appropriately called *degree of genetic determination* is found by dividing the total population variance into the genotypic variance. The resulting statistic tells us what proportion of the total variance of the population is determined by genotypic differences within the population.

Heritability in the narrow sense is a more useful concept. To calculate heritability in the narrow sense, we separate from the component of genetic variance that proportion which represents additive genetic variance and divide additive genetic variance by the total variance of the population. Additive variance is the variance of breeding values of the individuals in the population. The breeding value of an individual can be determined by mating him with several other members of the population and noting how the mean response of the resulting progeny differs from the mean response of the entire population. According to Falconer, additive variance ''is the important component since it is the chief cause of resemblance between relatives and therefore the chief determinant of the observable genetic properties of the population and *of the response of the population to selection*'' (1960, p. 135, italics added).

Several additional points should be made about the concept of heritability. First, note that an estimate of heritability is dependent upon the total variance of the population. Since the variance of the population includes components due to environmental differences, the heritability estimate will change if these are increased or decreased. Indeed, considering the degree of genetic determination, if environmental variation were reduced to zero, the degree of genetic determination would equal 100%. As Fuller and Thompson have written, ''Heritability is not an attribute of a trait with a fixed value which more and more refined methods will define with greater and greater precision, but *a characteristic of a population* with respect to a particular trait'' (1960, p. 64, italics added).

Second, the concept of additive variance used to define heritability in the narrow sense does not assume additive gene action. ''Additive variance can arise from genes with any degree of dominance . . .'' (Falconer, 1960, p. 138).

Third, characters closely related to reproductive fitness such as conception rate, litter size, viability, age at puberty, etc., usually result in low estimates of heritability in the narrow sense (Falconer, 1960, pp. 167–168). In other words, traits related to reproductive fitness would be difficult to change by artificial selection. Most animals in the population have the same breeding value. Additive genetic variance has been largely exhausted by natural selection.

3. The third question concerns the mode of inheritance of the behavior. Mode of inheritance is determined by examination of some measure of central tendency such as the mean. Three major modes of inheritance are possible: (1) intermediate inheritance, where the F_1 scores near the middle of the range set by the two parental values; (2) partial or complete dominance, where the F_1 resembles one of the parental strains; and (3) overdominance, where the F_1 scores outside the range set by the two parental strains. Mode of inheritance, theo-

retically at least, can lead to hypotheses regarding the survival value of particular behavioral traits. Because genetic variation in traits related to fitness is due mainly to dominance components, such traits are particularly subject to inbreeding depression. Conversely, heterosis or hybrid vigor occurs for such traits when inbred strains are crossed. Neutral traits, not directly related to fitness, do not reveal heterosis. The theory behind these statements has been summarized by Bruell (1967).

Now these are only three of many different questions that can be answered by behavior–genetic research. Indeed, they may not even be among the most important questions. I have argued elsewhere (McGill, 1969) that the primary contributions of behavior genetics may be in the areas of causation and development rather than evolution and function. The three questions are listed here because they are important to the current topic.*

We turn now to an examination of some representative experiments involving laboratory populations of house mice and note some of the answers to our three questions.

1. Do genotypic differences in sexual responses exist? Several years ago (McGill, 1962) I examined the sexual behavior of three different inbred strains of mice. I divided the male mating pattern into as many quantifiable elements as seemed to make sense and then compared the three strains on these elements of behavior. Analysis showed that it was the rule, rather than the exception, to find statistically significant differences in the various comparisons. Since inbred laboratory populations were originally derived from wild populations, we may hypothesize that considerable genetic variability, and probably behavioral variability, exists among wild members of the species.

2. The second question concerns heritabilities. Table IV presents some sample heritabilities that we have established for certain measures of sexual behavior in certain populations under particular testing circumstances. The method was that of a classical Mendelian analysis involving two inbred parent strains, reciprocal F_1's, F_2's, and backcross groups (McGill, 1970a). Keeping in mind the fact that heritability is a characteristic of a population and that the estimate of heritability is completely dependent upon the degree to

---

*And because their inclusion in Table I completes a mnemonic device that I have found extremely useful in teaching undergraduates. Notice that the initial letters of the nine ꙍords of the table are in alphabetical order from A through I.

Table IV. Estimates of Heritability for Certain Measures of Sexual Behavior in House Mice

	Broad (°GD)	Narrow (h^2)
Mount latency	.19	0
Intromission latency (log)	.13	.13
Number of mounts	.42	.39
Number of intromissions	.26	.26
Thrusts per intromission (log)	.41	.10
Total number of thrusts	.30	.30
Ejaculation latency	.18	.18
Interval between ejaculations	.70	.60

which we were successful in controlling environmental factors, the numbers in Table IV must be viewed as anything but absolute. Furthermore, for two measures we were required to transform the scores to logarithms in order to meet certain scaling assumptions. In other cases, no scaler transformation was successful in meeting these assumptions. However, transformation of the data to logarithms or to square roots did not greatly alter estimates of heritability in most instances.

Despite the limitations just mentioned, the heritabilities presented in Table IV are instructive. Note first that in many cases estimates of heritability, both broad and narrow, are fairly substantial, indicating that genotypic differences contribute importantly to variation in the trait. Second, for most measures much or all of the genotypic variation is due to additive genetic variation. Theoretically, this would indicate that these traits are not directly related to reproductive fitness and that the raw material for selection, either natural or artificial, exists in the population. But there is a problem with this interpretation. In order for biometrical genetics to be used to analyze a population's evolution, the animal groups studied must have been derived from the same base population and must represent an unbiased sample of the genes from that base population (Bruell, 1967; Maxson, 1973; Roberts, 1967). Unfortunately, currently available inbred strains do no meet either of these assumptions, and their usefulness in answering questions regarding the evolution and adaptive value of traits is minimal.

3. The third question concerns the mode of inheritance of the behavior. In our examination of modes of inheritance of various

quantitative elements of sexual behavior, we have found that some traits exhibit intermediate inheritance, some traits partial or complete dominance, and some few traits are inherited in a heterotic fashion (McGill & Blight, 1963).

An interesting example of heterotic inheritance is that of retention of the ejaculatory reflex following castration. Tucker and I (1964) first reported this phenomenon. We castrated male mice of the C57BL/6 and DBA/2 inbred strains and males resulting from a cross between C57BL/6 females and DBA/2 males. The hybrid males retained the ejaculatory reflex after castration significantly longer than males of either inbred strain. In a more recent experiment (McGill & Haynes, 1973), degree of heterozygosity was used as an independent variable and retention of the ejaculatory reflex after castration as a dependent variable. We found a significant positive correlation of .77 between the variables.

Now it is obvious that retention of the ejaculatory reflex after removal of the gonads is a trait that has never been subjected to selective pressure. While it is risky to generalize about mode of inheritance on the basis of a single cross (and we have evidence that not all F_1 groups respond as BDF_1 males do: McGill & Manning, 1976), nevertheless, the extreme hybrid vigor shown for this character in this cross is intriguing. It may indicate the existence of some other trait that is correlated with retention of the ejaculatory reflex after castration. The discovery of such a character might aid in identifying the physiological mechanisms underlying individual differences in post-castration sexual performance, a problem that has baffled investigators for many years (Young, 1961).

To reiterate a point made above, it may well be that the major contributions of behavior genetics will lie not in the establishment of such items as heritability and mode of inheritance but rather in the formulation of testable hypotheses concerning physiological differences underlying behavioral differences. For example, Bentley (1971) has made remarkable progress in correlating genetic differences with differences in a neuronal network responsible for the production of the species-specific song of crickets. Working with the same species, Hoy and Paul (1973) provided evidence for genetic control of song reception. Similar data have been gathered for cricket frogs by Capranica, Frishkoff and Nevo (1973). Ideally, mammalian behavior genetics will some day reach such levels of analysis.

As noted, in order for behavior–genetic research to have the greatest meaning for evolutionary biology, results should be obtained from wild populations. During a sabbatical year at the University of Edinburgh, I had an opportunity to apply some of the behavior–genetic techniques I have described to two subspecies of a naturally occurring rodent.

The bank vole, *Clethrionomys glareolus*, occurs throughout the mainland of Great Britain and also on certain islands off the West Coast. It is estimated that the mainland and island populations have been geographically separated for perhaps 7000 to 9000 years. Some of the island groups are sufficiently different morphologically and behaviorally to have been classified as distinct species (Steven, 1953). More recent work has indicated that the groups are capable of producing fertile hybrids and are more appropriately considered subspecies rather than separate species. However, since Godfrey (1958) has found some evidence for sexual selection, hybrid inviability, and hybrid breakdown, the subspecies may be in the process of speciating.

Two of these subspecies are maintained in the Department of Zoology at Edinburgh by Dr. John Godfrey. These are the mainland form, *Clethrionomys glareolus britannicus*, and *Clethrionomys glareolus skomerensis* from the island of Skomer, located 1½ miles off the Welsh coast. Skomer is a treeless island about 1¼ mile long and a mile wide. Skomer voles are larger than the mainland form, 25 grams compared to 18 grams adult weight, and they are sandy rather than reddish-brown in color. They exhibit remarkable tameness to man both when caught in the wild and when reared in the laboratory. This trait may be due to the fact that there are no mammalian predators present on Skomer. In the laboratory, Skomer voles are docile in novel situations and slow to initiate retrieval of young placed outside the nest (Alder, 1972, 1975).

We examined the sexual behavior of males of these two subspecies and of reciprocal crosses between them (Alder, Godfrey, McGill, & Watt, in prep.). The island males were from the third and fourth generations of animals originally caught on Skomer, while the mainland males were from the third and fourth generations of animals caught within 50 miles of Edinburgh.

The pattern of sexual behavior for the species is indicated in Table III. Intromissions consist of a single thrust, multiple intromissions occur, and there are multiple ejaculations prior to sexual exhaustion.

Table V. Mount Latency: Bank
Voles[a]

	br ♂	sk ♂
br ♀	97	28
sk ♀	308	159

[a] Arrows indicate statistically significant differences between pairs.

We found that the subspecies of *Clethrionomys* are similar to inbred strains of mice in that many statistically significant differences in quantitative elements of sexual behavior exist. The smaller mainland subspecies showed a more active pattern of mating than did the Skomer voles. Mainland males took less time to initiate mating, exhibited more intromissions and mounts before ejaculation, had shorter intervals between intromissions, and resumed mating sooner after ejaculation than did the island males. Considering the hybrids, patterns of intermediate inheritance and partial and complete dominance were noted. There were no clear instances of overdominance. The experimental design did not permit estimates of heritabilities.

These results refer to pairs of animals of the same subspecies. To study the contribution of each sex to the behavioral scores, we observed further matings where males of one subspecies were paired with females of the other subspecies. The results were instructive. Table V shows an example for mount latency, defined as the number of seconds from pairing to the first mount. Mainland pairs had significantly shorter mount latencies than Skomer pairs. Was this difference due to the males, or the females, or to the particular combination of males and females? Table V indicates that the latter was the case. The interactions are complex and interesting. Skomer males had significantly shorter mount latencies than *britannicus* males when the subspecies of the female was held constant. On the other hand, mainland females elicited shorter mount latencies than did *skomerensis* females from both groups of males. Thus the shortest mount latencies occurred when Skomer males were paired with mainland females; the longest mount latencies were observed when mainland males were paired with Skomer females. The significant difference between mainland pairs and Skomer pairs is thus due primarily to the attractiveness of the smaller mainland female and not to more rapid arousal of mainland males.

Table VI indicates a different situation for intromission frequency, defined as the number of intromissions preceding the initial ejaculation. In this case, males performed a subspecies-characteristic number of intromissions regardless of the genotype of their female partners.

Table VII shows the results for the time interval from the initial ejaculation to the first mount of the second series. This measure is similar to intromission frequency in that the subspecies of the female appears to have had little effect.

Putting the results in terms of one theory of male rodent sexual behavior (Beach, 1956; McGill, 1965), female differences are very important for the male's arousal mechanism as measured by mount latency. The copulatory mechanism, indicated by the number of intromissions, and the ejaculatory mechanism, measured by the time

**Table VI. Intromission
Frequency: Bank Voles[a]**

	br ♂	sk ♂
br ♀	22	15
sk ♀	20	15

[a] Arrows indicate statistically significant differences between pairs.

**Table VII. Postejaculatory
Interval to Mount: Bank Voles[a]**

	br ♂	sk ♂
br ♀	641	1059
sk ♀	706	990

[a] Arrows indicate statistically significant differences between pairs.

from the ejaculation of the first series to the first mount of the second series, appear to be determined by the genotype of the male.

In summary, to this point, behavior–genetic investigations of quantitative aspects of copulation indicate wide genotypic differences both between inbred strains of mice and between naturally occurring populations of bank voles. Relatively few behavioral measures exhibit a heterotic mode of inheritance, particularly as regards the bank vole data. The presence of additive genetic variance for several measures in laboratory mice suggests that these traits might be altered by selection and is consistent with the finding of subspecific differences in bank voles. None of the quantitative differences studied are capable of functioning as reproductive isolating mechanisms, but the species differences in the mating patterns could do so. Are there other functions that the behavioral patterns might serve? This question is considered in the next section.

Functions of Norway-Rat and House-Mouse Mating Patterns: Induction of Luteal Activity and Sperm Transport

A few years ago Wilson, Adler, and LeBoeuf (1965) reported experiments concerning the function of preejaculatory intromissions for the laboratory Norway rat. Quite independently, using different techniques and a different species, Land and I (1967) reported similar experiments for laboratory house mice. Each of these papers initiated a series of reports on the problem, and some interesting species differences emerged.

Female Norway rats and female house mice normally have 4- to 5-day estrous cycles. They ovulate spontaneously, but, in the absence of copulation, a true luteal phase does not develop. If a mating is sterile, the female becomes progestational and does not return to estrus for 10 to 12 days. This condition is called *pseudopregnancy*. Thus one possible function of preejaculatory stimulation in both species might be the induction of the luteal phase of the estrous cycle.

We will consider the results for the laboratory rat first. Wilson, Adler, and LeBoeuf (1965) placed stimulus females, in hormonally induced estrus, with male rats. After the male had achieved several intromissions and was judged ready to ejaculate, an experimental female in natural estrus replaced the stimulus female. The male usually ejaculated with three or fewer intromissions with the second female. Nine females receiving three or fewer intromissions constituted their Low Intromission group. The High Intromission group consisted of females in natural estrus who had had four or more intromissions prior to the ejaculatory response of the male (mean = 9.4). The dependent variable was the number of females made pregnant by the treatment. The results showed that 9 out of 10 females of the High Intromission group had been impregnated while only 2 of 9 in the Low Intromission group became pregnant as a result of the mating.

In 1969 Adler reported a series of experiments on various aspects of the problem. First, he replicated the study of Wilson, Adler, and LeBoeuf, using a larger sample and a somewhat greater number of intromissions in the High Intromission group. Eighty-four percent of females receiving six or more intromissions were impregnated, whereas only 20% of females receiving three or fewer became pregnant.

As mentioned above, a sterile mating may result in a progestational state accompanied by a lack of receptivity for about 12 days. In his second experiment, Adler found that 82% of females in the Low Intromission group continued to exhibit behavioral estrous cycles. Thus a small number of intromissions with the ejaculatory reflex and deposition of the copulatory plug was largely ineffective in either impregnating the female or in making her pseudopregnant.

In the third experiment, females were given 14 or more intromissions, but the males were not permitted to ejaculate. Sperm were then injected into the uterine lumen. All four females so treated ceased to exhibit the behavioral estrous cycle and were found to have developed embryos in the uterus two weeks after treatment. This result indicates that the ejaculatory response and the copulatory plug are not necessary to trigger the luteal response in female rats.

In order further to validate that conclusion, Adler mated females in natural estrus to males who had been treated with guanethidine sulfate. This drug permits performance of the complete male mating behavior pattern including the ejaculatory reflex, but it blocks the ejaculation of any detectible substances. Eight of nine females so treated were made pseudopregnant, i.e., they stopped exhibiting behavioral estrous cycles. In the fifth experiment, females were given from 1 to

16 intromissions but no ejaculation. They were then tested for cessation of behavioral receptivity. With from 1 to 4 intromissions, 5% stopped cycling, while, with from 13 to 16 intromissions, 83% did so. These findings provide further evidence that the ejaculatory response and the copulatory plug are not necessary for the induction of the luteal phase of the cycle in female rats.

Other experiments were concerned with sperm transport and subsequent fertilization. They showed that in Low Intromission groups fertilization of the eggs did not occur, and, furthermore, the females did not have sperm in the uterus.

Thus the preejaculatory intromissions of the male rat serve to induce the progestational stage of the cycle and to facilitate transport of spermatozoa into the uterus. Chester and Zucker (1970) have reported results in essential agreement with those of Adler.

In an effort to delineate further the mechanism involved, Adler, Resko, and Goy (1970) measured the concentration of progesterone in plasma of female rats. Females in a High Intromission group had significantly higher quantities of progesterone in their blood than did females in a Low Intromission group or females receiving no copulatory stimulation. In addition, females in the High Intromission group showed increasing concentrations of the hormone over the first four days following sexual stimulation. Low Intromission females or control females did not show this increase in progesterone.

The similar series of investigations on the laboratory house mouse shows some interesting similarities and some even more interesting differences. Species differences in the pattern of intromission and ejaculation should be kept in mind. Male mice exhibit long intromissions with multiple thrusting (mean = about 30) and long ejaculatory patterns (mean = about 20 seconds). Therefore, the number of thrusts, rather than the number of intromissions per se, was used as an independent variable in experiments with the house mouse.

Land and McGill (1967) selected females in natural estrus on the basis of their response to sexually active males. One group received a low number of preejaculatory thrusts, the ejaculatory reflex, and the delivery of a copulatory plug. A second group received a variable number of preejaculatory thrusts, ranging up to 156, but did not experience the ejaculatory reflex or deposition of the copulatory plug. Two days after experimental treatment, females were placed into the home cages of sexually active indicator males. The females were examined daily for the presence of copulatory plugs deposited by the indicator males and for the birth of litters.

Females from the first group that received as few as 10 preejaculatory thrusts and the ejaculation became pregnant. Most females from the second group that received only preejaculatory thrusts returned to behavioral estrus four or five days after treatment. These results indicate that a large number of preejaculatory thrusts is neither necessary nor sufficient for the induction of the progestational phase in the female house mouse. The ejaculatory reflex and the presence of the copulatory plug are sufficient to induce luteal activity regardless of the number of preejaculatory thrusts.

In order to investigate further the functions of the copulatory plug, McGill, Corwin, and Harrison (1968) mated female mice to males that had been surgically rendered incapable of forming plugs by removal of the seminal vesicles and the coagulating glands. About half of the females were impregnated and the other half made pseudopregnant as a result of these matings. In both cases, luteal activity was induced in the female. Other experiments showed that castrated males who had lost the capacity to deposit copulatory plugs retained, for a time, the ability to induce luteal activity in females and, furthermore, that guanethidine-treated males who do not ejaculate a plug are capable of inducing pseudopregnancy in females (McGill, 1970b). The copulatory plug is thus unnecessary for the induction of the luteal phase of the cycle.

Since preejaculatory stimulation and the copulatory plug are unnecessary for the induction of luteal activity in female mice, McGill and Coughlin (1970) investigated events occurring during the lengthy ejaculatory reflex characteristic of the species. We separated male mice from estrous females at times ranging from 1 to 7 seconds following the shudder that marks the beginning of the reflex. Fifteen of 16 females separated from males within 2 seconds of the initiation of the ejaculatory reflex were unaffected by the copulation in that they were mated 4 or 5 days later by indicator males.* Most females left in contact with the male for 3 seconds or longer were either pregnant or pseudopregnant and thus had had luteal activity induced.

---

* In a replication of that aspect of the experiment (McGill, 1972), 13 of 15 females who experienced the entire copulatory pattern of the male up to the beginning of the ejaculatory reflex were mated by indicator males within 4 or 5 days of treatment.

Observation of the penises of males that had been removed from females indicated that external ejaculation of the copulatory plug began about 2 seconds after the beginning of the reflex. Ejaculation of the plug was preceded by a large increase in the diameter of the erect organ, particularly at the distal end, where it assumed a cuplike shape. Formation of this penile cup was followed by ejaculation of the copulatory plug. Lying on the plug, but not attached to it, was a yellow sperm-rich fluid. When the ejaculatory reflex is not interrupted and plugs are dissected from females, strands of the plug are observed to penetrate through the cervix. It is probable that this is the mechanism for sperm transport in the house mouse. The penile cup persisted for from 5 to 12 seconds and then regressed within a second or two.

Since neither preejaculatory stimulation nor the copulatory plug are necessary for the induction of luteal activity in female mice, it was hypothesized that mechanical stretching of the vagina and/or the cervix, which results from the penile cup, could serve as an effective stimulus to trigger the response. This hypothesis was tested by use of an artificial penis designed to provide stimulation similar to that of the penile cup (McGill, 1970b). Eighty-one percent of receptive females treated with the artificial penis for 30 seconds became pseudopregnant.

Table VIII summarizes the results for the two species and highlights some interesting differences. For rats, intromissions are both

Table VIII. Potential Stimuli and the Induction of Luteal Activity in Two Rodent Species

	Norway rat	House mouse
Intromissions	Necessary	Probably not necessary
	Sufficient	Not sufficient
Plug	Not necessary	Probably sufficient
		if so, then ↓
Penile cup	Not necessary (if present)	Probably not necessary
	Not sufficient	Sufficient

necessary and sufficient for the induction of the luteal response. For house mice, intromissions are not sufficient. Our data have not firmly ruled out the necessity of such stimulation, since in the experiment with the artificial penis the response to a sexually active male was used to assay receptivity. If preejaculatory stimulation is necessary, the amount required is minimal since females have been impregnated by a single intromission consisting of 10 thrusts and ejaculation.

For both the rat and the mouse, the copulatory plug is not necessary. The plug is not sufficient for the rat, but considering its greater relative size and the fact that it remains in the vagina of the female mouse for from 18 hours to 2 days (Asdell, 1964, p. 367), it is probably sufficient in this species. If the plug is sufficient, then the penile cup, while sufficient, is probably not necessary. The presence of the penile cup during ejaculation has not been demonstrated in the rat, although Hart (1968) describes similar changes in the penises of spinal preparations. Even if the penile cup is present in the rat, it is neither necessary nor sufficient for the induction of the progestational phase of the estrous cycle.

Norway rats and house mice have thus evolved different mechanisms by which the luteal phase of the estrous cycle is induced and by which sperm are transported into the uterus of the female. If future research shows that closely related, sympatric species employ different mechanisms, these could function in reproductive isolation by producing gametic or zygotic mortality.

The next section reviews a behavioral phenomenon discovered in the laboratory that probably functions in the natural situation, that is related to the induction of luteal activity, and that might serve as yet another hurdle to interspecific mating.

Functions of Norway-Rat and House-Mouse Mating Patterns: The Enforced-Interval Effect

As was noted above, two groups of investigators independently embarked on a search for the functions of the relatively lengthy pattern of sexual responses exhibited by male Norway rats and male house mice. Those working with rats have been successful: preejaculatory intromissions are necessary for both the induction of the progestational phase and for the transport of sperm into the uterus. In terms of the original purpose, the studies involving house mice have failed to find any function for preejaculatory stimulation. The final series of experiments to be described may indicate why this was the case.

In the first place, it has been known for a number of years that, in common with most other rodents for which information is available, a male house mouse that has been sexually inactive for some time takes much longer to mate and executes many more preejaculatory intromissions than one that has recently ejaculated (McGill, 1963).

Preejaculatory intromissions can be reduced in other ways. Male Norway rats normally achieve an intromission every 30 to 90 seconds and ejaculate after about 11 intromissions. If the female is removed between intromissions and then returned after a longer time than the male would normally have rested, the number of intromissions preceding ejaculation is reduced (Bermant, 1964; Larsson, 1959). When the intro:..issions are separated by from 2 to 3 minutes, the male rat ejaculates after only four or five intromissions. When the interval is 7 minutes or greater, the number of intromissions required for ejaculation becomes very large; some males failed to ejaculate after several dozen intromissions separated by such intervals (Bermant, 1967).

Since the *enforced-interval effect*, as it is called, appeared to be a second way in which the number of intromissions preceding ejaculation in male house mice might be reduced from that observed in ad libitum copulations, we initiated an investigation of the phenomenon with that species. Pilot studies soon indicated that the effect does indeed exist for male mice but that the time course is quite different from that seen for the rat. In our fourth attempt we finally succeeded in using intervals long enough to prevent ejaculation in male mice.

Sixteen groups of eight B6D2F$_1$ male mice were used. All males were tested in home cages of the usual shoebox size. One group was permitted to copulate without interference by the experimenter. For a second group, the female was removed immediately after each intromission and returned within 2 seconds. For the third group, the female was withheld for 45 seconds and, for the fourth group, 90 seconds. Females were withheld for 3 minutes following each intromission for Group 5. For succeeding groups, the interval was progressively doubled until Group 10 had intromissions separated by 96 minutes. We rounded that figure to 1½ hours and doubled it again to 3 hours for Group 11. Groups 12 through 16 had intromissions separated by progressive doubling of the interval. Thus Group 16 experienced one intromission every 4 days. Animals with intromissions separated by more than 1½ hours were given new females for each intromission and were allowed a maximum of 20 intromissions to achieve ejaculation.

As shown at the top of Figure 1, all males achieved ejaculation with interintromission intervals up to and including 3 hours. Seventy-five percent of the males at 6 hours and 12 hours ejaculated, while two of eight ejaculated when intromissions were separated by 24 hours. None of the animals in the 48-hour or 96-hour group exhibited an ejaculatory response.

The bottom curve of Figure 1 indicates the number of intromissions preceding ejaculation. The most efficient interintromission interval was 96 minutes, where the mean number of intromissions preceding ejaculation was only 7.5. This interval also had the smallest variance, as four of the eight animals required seven intromissions and the other four executed eight. The top curve shows the number of thrusts per intromission. The increase in intromission length with enforced intervals has also been electrically recorded for the single thrust that normally characterizes the response in the rat (Bermant, Anderson, & Parkinson, 1969; Carlsson & Larsson, 1962).

These results indicate that male mice are capable of storing copulatory stimulation for a much longer period of time than are male rats, perhaps for as long as 24 hours. In order to test this latter hypothesis, 40 sexually experienced B6D2F$_1$ males were divided into two groups on the basis of preliminary measurements of mating time. The first group of 20 animals was allowed to execute 100 thrusts with a receptive female, at which point the mating was interrupted. Twenty-four hours later, both groups of 20 males were given estrous females and permitted to copulate until the ejaculatory response occurred. The males that had experienced 100 thrust 24 hours earlier had significantly shorter matings, with significantly fewer intromissions and total thrusts, than did the control males. Experimental males executed significantly more thrusts per intromission than did the control animals. Table IX presents the findings.

The results show that the copulatory mechanism (Beach, 1956), as indicated by the number of intromissions, is sensitized by enforced intervals, in the case of the mouse for as long as 24 hours. Are other theoretical measures of sexual behavior affected? Figure 2 shows the

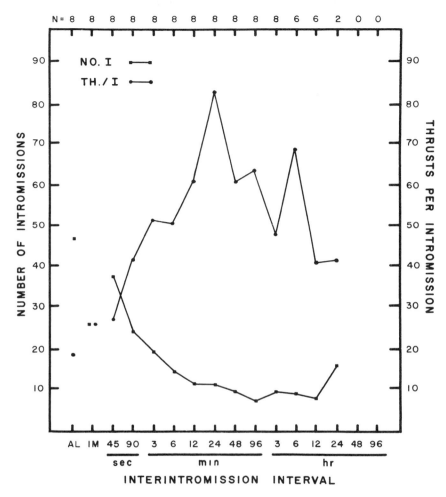

Figure 1. The effects of enforced intervals between intromissions on the number of intromissions preceding ejaculation (No. I) and on the number of thrusts per intromission (TH/I). AL = ad libitum copulation. IM = female removed and immediately replaced.

results for mount latency, a measure of the arousal mechanism (Beach, 1956). Mount latency was recorded each time the female was reintroduced following an enforced interval. The average of these mount latencies is expressed as a percentage of the original mount latency before any experimental manipulation took place. Mount laten-

Table IX. The Effects of an Incomplete Mating Twenty-Four Hours Prior to Testing on Certain Measures of Sexual Behavior in Male House Mice

	Median scores		
	100-Thrust group	Control group	p Value
Number of intromissions	21.0	40.5	.002
Ejaculation latency	1375 sec	2402 sec	.02
Total number of thrusts	568	890	.02
Thrusts per intromission	29.8	21.7	.02

cies for intervals up to and including 12 hours were less than 60% of those originally recorded. Beyond 12 hours, mount latency approaches the value measured for the initial response, and indeed we found no significant difference in mount latency for the 24-hour/100-thrust experiment described above. The results clearly show that the arousal mechanism is also sensitized by enforced intervals.

In 1965, I proposed a third theoretical mechanism, called the ejaculatory mechanism, to account for certain elements of sexual behavior in male rodents. It was presumed to discharge at ejaculation and to require some time to recover before mating resumed. Figure 3 indicates the length of the ejaculatory reflex for the various groups in the enforced-interval experiment. As the curve shows, increasing intervals resulted in a progressive increase in the length of the reflex. I reasoned that a longer ejaculatory reflex should reflect greater discharge of the ejaculatory mechanism. Greater discharge should then result in longer recovery time following the initial ejaculation. To test this hypothesis, 2 matched groups of 10 animals each were studied. One group copulated without interference, while the second group had a 90-minute enforced interval between intromissions. For both groups, 10 minutes after the initial ejaculation, a fresh estrous female was introduced into the male's cage. If mounting did not occur within 5 minutes, that female was removed, and 10 minutes later a new female was introduced for another 5-minute period. This procedure continued until the male achieved a second ejaculation. No intervals were enforced between intromissions in the recovery tests.

The results were directly opposite to my prediction, based on greater discharge of the ejaculatory mechanism. Eight of 10 enforced-interval males initiated and completed mating with the first female

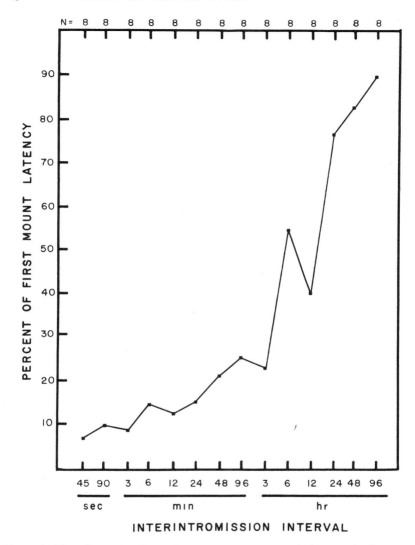

Figure 2. The effects of enforced intervals between intromissions on the latency to mount when the female was again available.

presented after the initial ejaculation. The other two did so with the second female. Only 1 of 10 control males mated with the first female. Median time between ejaculations for the enforced-interval group was 17.5 minutes, while that for the control animals was 32 minutes. The difference was significant at the .02 level by a Mann–

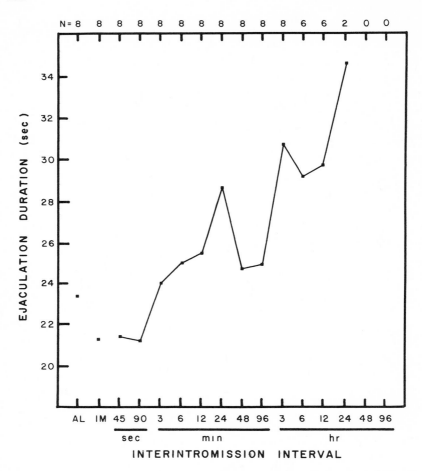

Figure 3. The effects of enforced intervals between intromissions on the length of the ejaculatory reflex.

Whitney U Test. In a second experiment, six animals were tested for recovery following enforced intervals of 6 hours between intromissions and were compared with a matched control group of six animals that mated without interference. All six of the enforced-interval males copulated to ejaculation with the first female offered following the initial mating; median recovery time between ejaculations was 13.75 minutes. None of the control animals mounted the first female; median recovery time for this group was 45.5 minutes. Enforced intervals, then, function to sensitize the ejaculatory mechanism so that recovery from the initial ejaculation occurs significantly sooner.

The results from our studies of the enforced-interval effect raise at least three questions. First, does the effect operate in the natural situation? Second, if it does, what are its functions? And third, why do Norway rats differ from house mice in the temporal aspects of the enforced-interval effect?

We have no direct data bearing on the occurrence of the enforced-interval effect in the natural situation, but we do have some indirect evidence. It is possible that wild females, operating in greater space and with more means of avoiding the advances of the male, might space intromissions further than are observed when the animals are mating in small, bare laboratory cages. As a first test of this notion, we compared the scores of 13 pairs of animals mating in aquaria measuring 13 inches in width by 35 inches in length with 13 pairs mating in laboratory cages (6½ inches by 11 inches). Mean interintromission interval in the aquaria was significantly longer than that observed in the small cages. There were significantly fewer intromissions preceding ejaculation for those pairs mating in the aquaria. These results support the hypothesis that intromissions may be more widely spaced for mice mating in a natural environment.

The second question involves the functions of the enforced-interval effect. Some possibilities are obvious: the male would mount receptive females sooner, have fewer intromissions, spend less time and energy in mating, and recover the capacity for a second ejaculation more rapidly.

Another possible function relates to the social organization of the species. Studies of the social organization of wild and laboratory populations of house mice in natural and experimental situations have been summarized by DeFries and McClearn (1972). Although there are gaps and some inconsistencies in the information, the best evidence indicates that mice live in small demes on small territories which they actively defend. There is almost always a dominant male and frequently two or three subordinate males who are probably sons of the dominant male. The dominant male may sire as many as 95% of all litters. The effective breeding size may be as few as four animals. There is evidence for a high degree of inbreeding and little genetic exchange with neighboring demes. Indeed, DeFries and McClearn note that an inbred mouse may not be as artificial as has sometimes been supposed. Furthermore, the success that has been experienced in producing inbred strains within this species may be due to the amount of inbreeding that had already occurred in the base populations.

As noted, the available evidence indicates that dominant males sire the vast majority of litters. Excluding the possibility of sexual selection on the part of the females (Yanai & McClearn, 1972, 1973a,b), there are reasons why a dominant male might be expected to be more successful in impregnating females. These stem from research described above indicating that engaging in sexual behavior results in sensitization of the arousal mechanism, the copulatory mechanism, and the ejaculatory mechanism.

If a male mouse has recently ejaculated, the next mating will be greatly abbreviated (McGill, 1963). Furthermore, if he has had an incomplete mating within 24 hours (and perhaps longer), his mating time will be reduced. Finally, when a female spaces intromissions, the male requires fewer intromissions and recovers sexual capacity rapidly. All of these factors operate to abbreviate mating time and preejaculatory stimulation *in sexually active males*. Conversely, sexual inactivity operates to lengthen arousal time, mating time, and recovery time after an ejaculation. Thus even though a subordinate male or nonterritorial male might initiate mating, his chances of reaching the ejaculatory threshold quickly would appear to be low.

The third question concerned the difference between Norway rats and house mice in the enforced-interval effect. In light of what is known about the induction of the luteal phase of the estrous cycle, this difference makes sense. A male rat must provide a certain amount of preejaculatory stimulation to induce the luteal response in the female and to insure the transport of spermatozoa through the cervix. This being the case, it would be adaptive for the male rat *not* to store copulatory information for very long, and indeed the effects of previous intromissions do not appear to accumulate when intromissions are spaced by 7 minutes or more.

For a female rat, on the other hand, storage of copulatory stimulation should be adaptive in the event that mating is interrupted. Edmonds, Zoloth, and Adler (1972) found that female rats are capable of storing copulatory stimulation for periods as long as one hour, and perhaps for considerably longer. The female rat needs copulatory stimulation and stores it. The male rat needs to provide sufficient stimulation to the female for induction of luteal activity and the transport of sperm and does not store copulatory stimulation for long periods.

The situation is quite different for the house mouse. In this species the female has no known need for copulatory stimulation beyond that experienced during the ejaculatory reflex. Preejaculatory stimulation is not necessary to transport sperm through the cervix; the copulatory plug achieves that end.* For the male mouse, rapid impregnation of estrous females would appear to be adaptive, and storage of copulatory stimulation results in rapid mating.

Concluding Remarks

To return again to the question of reproductive isolating mechanisms, it is obvious that the behavioral and physiological factors we have discussed might function in such a capacity in certain instances. It is probable, however, that these are secondary hurdles. It would seem much more adaptive for isolating mechanisms to operate prior to copulation to prevent useless wastage of energy and gametes. In a paper with the provocative title of "A Cry for the Liberation of the Female Rodent: Courtship and Copulation in *Rodentia*," Doty (1974) has reviewed evidence that reproductive isolation in this order is probably maintained through mate selection by females, primarily on the basis of olfactory cues. If this is the case, we need many more studies involving interspecies cross-fostering, such as that of Quadagno and Banks (1970).

When invited to contribute to this symposium volume, I was requested to consider the possible implications of behavior–genetic research to the understanding of reproductive isolation. I was not able to confine myself to that topic. However, even if the question is broadened to include all aspects of the evolution and survival value of behavior, I am not optimistic as to the value of behavior–genetic analysis in mammals, using currently available laboratory stocks. For

---

* The plug also functions, in most cases, to prevent double inseminations. In the rat, double inseminations appear to be reduced by inhibition of sperm transport when a second copulation follows within 15 minutes of the first (Adler & Zoloth, 1970). Thus, when double matings occur in house mice, the first male will probably sire the litter, while in the Norway rat the second male may do so.

example, Doty (1974) suggests that distinctive male and female odors may have been lost in the process of inbreeding.

Perhaps the best use of behavior–genetic experiments with laboratory animals might be as hypothesis-generating devices. These hypotheses are then to be tested using appropriate material, that is, animals from naturally occurring populations or animals carefully and properly derived from naturally occurring populations. This is one area where behavior geneticists who study *Drosophila,* birds, crickets, etc., have had a distinct advantage. It is time that those who work with mammals caught up.

But, at the same time, I do not wish to denigrate the proper use of currently available strains. They offer easily obtainable estimates of genetic and environmental contributions to variance, and they provide important material for the study of physiological differences underlying behavioral differences.

The research that I have reviewed regarding differences between Norway rats and house mice in the enforced-interval effect, storage of copulatory stimulation, induction of luteal activity, and mechanisms of transport of sperm through the cervix is remarkably consistent. Of course, many questions remain, primary of which might be why these two species have evolved quite different mechanisms to achieve the same ends. Considering only these two species, the answer cannot be that the differences function as reproductive isolating mechanisms. Aside from obvious differences in size, chromosome number, etc., rats tend to kill mice (O'Boyle, 1974). While this is a remarkably effective behavioral reproductive isolating mechanism, it does not serve to answer our question.

In light of the species differences between rats and mice, many questions arise concerning the variety of copulatory patterns shown in Table III. In which of these species must the female be induced to ovulate? In which must the luteal phase of the cylcle be induced? What is the mechanism by which such induction takes place? What is the mechanism for sperm transport? Does the same pattern always serve the same function? What is the social structure of the species, and how do differences in social structure correlate with differences in physiological adaptations? How frequently and in what social context do sympatric species interact? Do sympatric species tend to have different physiological mechanisms associated with different copulatory patterns?

This series of questions can be answered only by a combination of laboratory experiments and observations of the animals in their natural environments.

ACKNOWLEDGMENTS

Research undertaken by the author and described in this chapter was supported by Research Grant 07495 from the Institute of General Medical Sciences, U.S. Public Health Service. I thank Beth Alder, Don Dewsbury, Lee Drickamer, John Godfrey, Aubrey Manning, Steve Maxson, and Ben Sachs for helpful comments on various drafts of the chapter.

A personal note: I feel privileged to have taken part in this first symposium dedicated to the memory of Daniel S. Lehrman. While most of the other contributors knew Danny longer and had more direct contact with him than I did, I am perhaps in a better position to judge one of his less well-known contributions to animal behavior. That was his willingness to pay repeated visits to undergraduate institutions to instruct, entertain, and inspire bright young minds. In the light of his many other obligations, it would have been understandable had he rejected such invitations, but in my experience he never did. Danny was scheduled for his third visit to Williams at the time of his death, and I know from invitations and posters that crossed my desk that he visited many other colleges in the East. The impact of these efforts on both those undergraduates who entered the sciences and those who will someday make decisions that affect science could conceivably be as important as his other, more visible, contributions.

References

Adler, N. T. Effects of the male's copulatory behavior on successful pregnancy of the female rat. *Journal of Comparative and Physiological Psychology*, 1969, *69*, 613–622.

Adler, N. T., & Zoloth, S. R. Copulatory behavior can inhibit pregnancy in female rats. *Science*, 1970, *168*, 1480–1482.

Adler, N. T. Resko, J. A., & Goy, R. W. The effect of copulatory behavior on hormonal change in the female rat prior to implantation. *Physiology and Behavior*, 1970, *5*, 1003–007.

Adler, E. M. The behavior of two races of the bank vole, *Clethrionomys glareolus* (Schreber, 1780). Ph.D. Thesis, University of Edinburgh, 1972.

Alder, E. M. Genetic and maternal influences on docility in the skomer role, *Clethrionomys glareolus skomerensis. Behavioral Biology*, 1975, *13*, 251–255.

Alder, E. M., Godfrey, J., McGill, T. E., & Watt, K. R. Sexual behaviour of two sub-species of the bank vole (*Clethrionomys glareolus*). In preparation.

Alexander, R. D. Sound communication in orthoptera and cicadidae. In W. E. Lanyon & W. N. Tavolga (Eds.), *Animal sounds and communication*. Washington, D.C.: American Institute of Biological Sciences, 1960. Pp. 33–92.

Alexander, R. D. Evolutionary change in cricket acoustical communication. *Evolution*, 1962, *16*, 443–467.

Arbib, R. Hail, great-tailed grackle! Baltimore oriole, farewell! *Audubon*, 1973, *75*(6), 36–39.

Asdell, S. A. *Patterns of mammalian reproduction*. (2nd ed.). Ithaca: Cornell University Press, 1964.

Bastock, M. *Courtship: An ethological study*. Chicago: Aldine, 1967.

Beach, F. A. Characteristics of masculine "sex drive." In M. R. Jones (Ed.), *Nebraska symposium on motivation*. Lincoln, Nebraska: University of Nebraska Press, 1956. Pp. 1–38.

Beach, F. A., & Jordon, A. Sexual exhaustion and recovery in the male rat. *Quarterly Journal of Experimental Psychology*, 1956, *8*, 121–133.

Bentley, D. R. Genetic control of an insect neuronal network. *Science*, 1971. *174*, 1139–1141.

Bermant, G. Effects of single and multiple enforced intercopulatory intervals on the sexual behavior of male rats. *Journal of Comparative and Physiological Psychology*, 1964, *57*, 398–403.

Bermant, G. Copulation in rats. *Psychology Today*, 1967, *1*(7), 53–60.

Bermant, G., Anderson, L., & Parkinson, S. R. Copulation in rats: Relations among intromission duration, frequency, and pacing. *Psychonomic Science*, 1969, *17*, 293–294.

Blair, F. Isolating mechanisms and interspecies interactions in auran amphibians. *Quarterly Review of Biology*, 1964, *39*, 334–344.

Bruell, J. H. Behavioral heterosis. In J. Hirsch (Ed.), *Behavior–genetic analysis*. New York: McGraw–Hill, 1967. Pp. 270–286.

Capranica, R. C., Frishkoff, L. S., & Nevo, E. Encoding of geographic dialects in the auditory system of the cricket frog. *Science*, 1973, *182*, 1272–1275.

Carlsson, S. G., & Larsson, K. Intromission frequency and intromission duration in the male rat mating behavior. *Scandinavian Journal of Psychology*, 1962, *3*, 189–191.

Chester, R. V., & Zucker, I. Influence of male copulatory behavior on sperm transport, pregnancy and pseudopregnancy in female rats. *Physiology and Behavior*, 1970, *5*, 35–43.

Committee. Thirty-second supplement to the American Ornithologists' Union check list of North American birds. *The Auk*, 1973, *90*, 411–419.

DeFries, J. C., & McClearn, G. E. Behavioral genetics and the fine structure of mouse populations: A study in microevolution. In T. Dobzhansky, M. K.

Hecht, & W. C. Steere (Eds.), *Evolutionary biology.* Vol. 5. New York: Appleton-Century-Crofts, 1972. Pp. 279–291.

Dewsbury, D. A. Patterns of copulatory behavior in male mammals. *Quarterly Review of Biology*, 1972, *47*, 1–33.

Doty, R. L. A cry for the liberation of the female rodent: Courtship and copulation in *Rodentia. Psychological Bulletin*, 1974, *81*, 159–172.

Edmonds, S., Zoloth, S. R., & Adler, N. T. Storage of copulatory stimulation in the female rat. *Physiology and Behavior*, 1972, *8*, 161–164.

Ehrman, L. The genetics of sexual isolation in *Drosophila paulistorum. Genetics*, 1961, *46*, 1025–1038.

Ehrman, L. Hybrid sterility as an isolating mechanism in the genus *Drosophila. Quarterly Review of Biology*, 1962, *37*, 279–302.

Ehrman, L. Direct observation of sexual isolation between allopatric and sympatric strains of the different *Drosophila paulistorum* races. *Evolution*, 1965, *19*, 459–464.

Falconer, D. S. *Introduction to quantitative genetics.* New York: Ronald Press, 1960.

Fuller, J. L., & Thompson, W. R. *Behavior genetics.* New York: Wiley, 1960.

Godfrey, J. The origin of sexual isolation between bank voles. *Proceedings of the Royal Physical Society of Edinburgh*, 1958, *27*, 47–55.

Hart, B. L. Sexual reflexes and mating behavior in the male rat. *Journal of Comparative and Physiological Psychology*, 1968, *65*, 453–460.

Hinde, R. A. Behaviour and speciation in birds and lower vertebrates. *Biological Reviews*, 1959, *34*, 85–128.

Hinde, R. A., & Stevenson–Hinde, J. *Constraints on learning.* London: Academic Press, 1973.

Hoy, R. R., & Paul, R. C. Genetic control of song specificity in crickets. *Science*, 1973, *180*, 82–83.

Land, R. B., & McGill, T. E. The effects of the mating pattern of the mouse on the formation of corpora lutea. *Journal of Reproduction and Fertility*, 1967, *13*, 121–125.

Larsson, K. The effect of restraint upon copulatory behavior in the rat. *Animal Behavior*, 1959, *7*, 23–25.

Lehrman, D. S. Semantic and conceptual issues in the nature–nurture problem. In L. R. Aronson, E. Tobach, D. S. Lehrman, & J. S. Rosenblatt (Eds.), *Development and evolution of behavior: Essays in memory of T. C. Schneirla.* San Francisco: Freeman, 1970. Pp. 17–52.

Manning, A. The sexual isolation between *Drosophila melanogaster* and *Drosophila simulans. Animal Behaviour*, 1959, *7*, 60–65.

Manning, A. *Drosophila* and the evolution of behaviour. *Viewpoints in Biology*, 1965, *4*, 125–169.

Manning, A. Evolution of behavior. In J. L. McGaugh (Ed.), *Psychobiology: Behavior from a biological perspective.* New York: Academic Press, 1971. Pp. 1–52.

Maxson, S. C. Behavioral adaptations and biometrical genetics. *American Psychologist*, 1973, *28*, 268–269.

Mayr, E. *Animal species and evolution.* Cambridge: Harvard University Press, 1963.

Mayr, E. *Populations, species, and evolution*. Cambridge: Harvard University Press, 1970.

McGill, T. E. Sexual behavior in three inbred strains of mice. *Behaviour*, 1962, *19*, 341–350.

McGill, T. E. Sexual behavior of the mouse after long-term and short-term postejaculatory recovery periods. *Journal of Genetic Psychology*, 1963, *103*, 53–57.

McGill, T. E. Studies of the sexual behavior of male laboratory mice: Effects of genotype, recovery of sex drive, and theory. In F. A. Beach (Ed.), *Sex and behavior*. New York: Wiley, 1965. Pp. 76–88.

McGill, T. E. Behavior genetic analysis in the mouse. Paper read at the Plenary Session on Behavior Genetics at the Eleventh International Ethological Conference, Rennes, France, 1969.

McGill, T. E. Genetic analysis of male sexual behavior. In G. Lindzey & D. D. Thiessen (Eds.), *Contributions to behavior-genetic analysis: The mouse as a prototype*. New York: Appleton-Century-Crofts, 1970a. Pp. 57–88.

McGill, T. E. Induction of luteal activity in female house mice. *Hormones and Behavior*, 1970b, *1*, 211–222.

McGill, T. E. Preejaculatory stimulation does not induce luteal activity in the mouse *Mus musculus*. *Hormones and Behavior*, 1972, *3*, 83–85.

McGill, T. E., & Blight, W. C. The sexual behaviour of hybrid male mice compared with the sexual behaviour of males of the inbred parent strains. *Animal Behaviour*, 1963, *11*, 480–483.

McGill, T. E., & Coughlin, R. C. Ejaculatory reflex and luteal activity induction in *Mus musculus*. *Journal of Reproduction and Fertility*, 1970, *21*, 215–220.

McGill, T. E., & Haynes, C. M. Heterozygosity and retention of ejaculatory reflex after castration in male mice. *Journal of Comparative and Physiological Psychology*, 1973, *84*, 423–429.

McGill, T. E., & Manning, A. Genotype and retention of the ejaculatory reflex in castrated male mice. *Animal Behavior*, 1976, *24*, 507–518.

McGill, T. E., & Tucker, G. R. Genotype and sex drive in intact and in castrated male mice. *Science*, 1964, *145*, 514–515.

McGill, T. E., Corwin, D. M., & Harrison, D. T. Copulatory plug does not induce luteal activity in the mouse *Mus musculus*. *Journal of Reproduction and Fertility*, 1968, *15*, 149–151.

O'Boyle, M. Rats and mice together: The predatory nature of the rat's mouse-killing response. *Psychological Bulletin*, 1974, *81*, 261–269.

Parsons, P. A. *Behavioral and ecological genetics: A study in Drosophila*. Oxford: Clarendon Press, 1973.

Quadagno, D. M., & Banks, E. M. The effect of reciprocal crossfostering on the behavior of two species of rodents, *Mus musculus* and *Baiomys taylori ater*. *Animal Behaviour*, 1970, *18*, 379–390.

Roberts, R. C. Some evolutionary implications of behavior. *Canadian Journal of Genetics and Cytology*, 1967, *9*, 419–435.

Seligman, M., & Hager, J. *Biological boundaries of learning*. New York: Appleton-Century-Crofts, 1972.

Spiess, E. B. Mating propensity in its genetic basis in *Drosophila*. In M. K. Hecht & W. C. Steere (Eds.), *Essays in evolution and genetics in honor of Theodosius Dobzhansky: A supplement to evolutionary biology,* New York: Appleton-Century-Crofts, 1970. Pp. 315–379.

Spieth, H. T. Mating behavior within the genus *Drosophila* (Diptora). *Bulletin of the American Museum of Natural History*, 1952, *99*, 401–474.

Steven, D. M. Recent evolution in the genus *Clethrionomys*. In *Symposium of the society for experimental biology: Evolution*. New York: Academic Press, 1953. Pp. 310–319.

Tinbergen, N. On aims and methods of ethology. *Zeitschrift für Tierpsychologie*, 1963, *20*, 410–433.

Tinbergen, N., Broekhuysen, G. J., Feekes, F., Houghton, J. C. W., Kruuk, H., & Szulc, E. Egg shell removal by the Blackheaded Gull, *Larus ridibundus L.*; A behaviour component of camouflage. *Behaviour*, 1962, *19*, 74–117.

Tinbergen, N., Kruuk, H., & Paillette, M. Egg shell removal by the Black-headed Gull. *Bird Study*, 1962, *9*, 123–131.

Wilson, J. R., Adler, N. T., & LeBoeuf, B. The effects of intromission frequency on successful pregnancy in the female rat. *Proceedings of the National Academy of Sciences*, 1965, *53*, 1392–1395.

Yanai, J., & McClearn, G. E. Assortative mating in mice. I. Female mating preference. *Behavior Genetics*, 1972, *2*, 173–183.

Yanai, J., & McClearn, G. E. Assortative mating in mice. II. Strain differences in female mating preference, male preference, and the question of possible sexual selection. *Behavior Genetics*, 1973a, *3*, 65–74.

Yanai, J., & McClearn, G. E. Assortative mating in mice. III. Genetic determination of female mating preference. *Behavior Genetics*, 1973b, *3*, 75–84.

Young, W. C. The hormones and mating behavior. In W. C. Young (Ed.), *Sex and internal secretions*. Baltimore: Williams and Wilkins, 1961. Pp. 1173–1239.

5

Evolutionary Aspects of Neuroendocrine Control Processes

Berta Scharrer

The orientation of this symposium should bring into focus not only the evolution and diversification of behavioral patterns but also the basic principles of their governance. In essence, the adjustment of an animal's behavior to variable conditions, ensuring the survival of the individual and the species, requires an apparatus for the reception, sorting out, and integration of relevant signals and the activation of effective response mechanisms. In simple as well as elaborate systems, the organ responsible for both the design of strategy and the execution of appropriate behavioral actions is the brain.

Reproductive behavior, the special theme of this program, involves long-term adjustments and periodic changes and thus depends on the complex interaction of environmental and internal cues. Among the latter, those provided by the endocrine system play a decisive role. The elucidation of these phenomena requires an understanding of the adaptive mechanisms available for effective communication between the nervous and the endocrine systems, each of which operates in its own characteristic way. The following discussion with its concentration on the evolutionary aspects of neuroendocrine interaction is intended as a contribution toward this end.

Berta Scharrer · Department of Anatomy, Albert Einstein College of Medicine, Bronx, New York 10461.

Principles of Neuroendocrine Integration

The pathways and mechanisms available for extero- and interoceptive sensory input to neural centers responsible for behavioral directives need no discussion. Afferent signals dispatched by the endocrine apparatus depend upon a capacity on the part of at least certain neurons to "sense" changes in blood hormone titers, i.e., to respond to a form of communication which differs from their own. The existence of such specially attuned neurons can be deduced, for example, from the demonstration that differential steroid binding (Meyer, 1973; Zigmond, Nottebohm, & Pfaff, 1973; Kato, Atsumi, & Inaba, 1974) and conversion of androgens to estrogens (Weisz & Gibbs, 1974) are restricted to specific areas of the central nervous system (hypothalamus, amygdala). The isolation, at successive developmental stages, of the special macromolecules (proteins) that serve as estrogen receptors in the anterior hypothalamus shows a good correlation between the maturation of this uptake mechanism and the animals' growing capacity to make use of afferent hormonal stimuli.

Such signals, passed on from these localized receptor sites to other neurons, become part of a body of information which, in turn, elicits two types of response. The first, and most important in the context of the present discussion, is a clearly demonstrable effect of circulating hormones, e.g., sex steroids, on behavior.

The second type of neural response to afferent hormonal signals is of more indirect concern to the management of behavior in that it elicits changes in blood hormone levels, a prerequisite for the aforementioned response. The initial and crucial step in this control mechanism is the dispatch of directives by the brain to the endocrine apparatus, i.e., to the vertebrate pituitary, or to analogous invertebrate organs.

The versatile manner in which this neuroendocrine axis operates has been discussed on several occasions (see E. & B. Scharrer, 1963 to 1976). Suffice it to say here that its efficacy depends on highly specialized (*neurosecretory*) neurons in which the capacity to produce chemical messengers has been developed to such a degree that it has become their primary activity.

The reason for their existence seems to be that neurohormones reaching endocrine effector cells, either via the general or a more directed (portal) circulatory route, are ideally suited for producing simultaneous and sustained stimulation of these cells.

Evolutionary Levels of Neuroendocrine Control of Behavior

Neuroendocrine activities determining reproductive behavior can be demonstrated not only in vertebrates but also in arthropods, mollusks, echinoderms, and annelids. These phenomena encompass a wide range of behavior patterns of varying complexity pertaining to courtship, mating, and parental care; periodic maturation and release of gametes; sex reversal, etc.; and the specific control mechanisms underlying each of these activities can be expected to be equally diversified. Nevertheless, there are common denominators. For example, a recurring feature, observed from worms to mammals, is the participation of inhibitory stimuli, determining the onset and duration of sexual activity and the concomitant modulations in behavioral patterns.

The examples selected for the discussion which follows should suffice to illustrate the salient features on a comparative basis. The periodic sequence of reproductive events characteristic of mammals and other *vertebrates* (dealt with in several contributions to this symposium) is programmed by the neuroendocrine apparatus discussed in the preceding section. Little needs to be added here, except for some remarks on the so-called releasing or hypophysiotropic factors by means of which stimulatory and inhibitory hypothalamic directives are conveyed to specific effector cells of the pituitary gland. Derived from neurosecretory cells and transported by a semiprivate vascular channel, these small peptides qualify as neurohormones. In some lower vertebrates, this circulatory (portal) route is bypassed, and the release sites of the respective neurosecretory mediators may approach their cells of destination as closely as those of regular neurotransmitters.

A point of particular interest is that some of these special neurochemical mediators seem to be capable of addressing themselves to neurons as well as to glandular cells. Tests with synthetically produced hypophysiotropins have revealed such effects that are not mediated by the pituitary. One of these mimics the action of L-dopa on dopaminergic neurons of the brain; another effect may be that of an antidepressant (Plotnikoff, Kastin, Anderson, & Schally, 1972; Ehrensing & Kastin, 1974; Horst & Spirt, 1974). The first direct neurochemical evidence that such a hypothalamic neurohormone affects the synthesis of the neurotransmitter dopamine in striatal neurons of the rat was provided by Friedman, Friedman, and Gershon (1973).

An even more pertinent example is the facilitation by luteinizing hormone-releasing factor (LRF) of lordosis behavior in estrogen-primed, hypophysectomized, ovariectomized rats (Pfaff, 1973). This neurosecretory mediator may reach neurons concerned with lordosis behavior via collaterals of LRF-producing neurons whose major axonal projection is to the median eminence (hypothalamo-infundibular tract). In the guinea pig, the existence of such extrahypophysial neurosecretory pathways (certain axons or axon collaterals) has been demonstrated by means of an indirect immunofluorescence reaction (Barry, Dubois, & Poulain, 1973; see also Zimmerman, 1976). Furthermore, there is increasing evidence for a role of vasopressin in the maintenance of active and passive behavior (De Wied, van Wimersma Greidanus, Bohus, Urban, & Gispen, 1976).

The intricacies of neuroendocrine events involved in the reproductive behavior of some of the higher *invertebrates* are as complex as those of vertebrates. The behavioral components observed in arthropods show considerable variability and species-specificity. Many forms display an elaborately programmed sequence of courtship, insemination, egg development, and oviposition (or parturition). These events are closely correlated with endogenous developmental processes, including the molt cycle. Adjustments to unfavorable environmental conditions are accomplished by quiescent phases (diapause). Some of the behavior patterns show relationships with diurnal rhythmicity.

The neuroendocrine apparatus carrying out the necessary directives in *insects* is analogous to the hypothalamic–hypophysial system of the vertebrates and makes use of hormonal factors from neural as well as nonneural sources. Neurosecretory messengers are provided by several ganglia and by the corpus cardiacum, which also serves as a neurohemal organ for the storage and release of cerebral neurohormones. The corpus allatum, an endocrine gland which is under stimulatory and inhibitory neural control (Figure 1), serves various func-

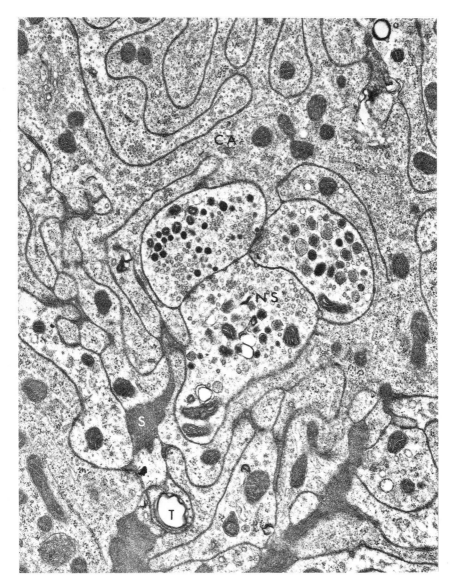

Figure 1. Electron micrograph of parts of interdigitating corpus allatum cells (CA), "innervated" by a group of three neurosecretory fibers (NS) of protocerebral origin. S, stromal partition; T, tracheole. Adult male of *Arphia pseudonietana*, Acrididae. × 20,750. (Specimen courtesy of Dr. S. N. Visscher, Montana State University.)

tions, but sex hormones seem to play a minor role. An active principle comparable to gonadal vertebrate hormones demonstrated thus far is the male hormone of the beetle *Lampyris*, which originates in the apical testicular tissue, controls the development of male sexual characters, and is governed by a hormone derived from neurosecretory brain cells (Naisse, 1969).

Studies in moths (Truman, 1973a; Truman & Riddiford, 1974) have shown that a sequence of hormonal signals to selected areas of the central nervous system brings about a sequence of events from eclosion to oviposition. The emergence of the adult insect, programmed for a specific period of the day by a circadian clock sensitive to photoperiod, results from the action of an "eclosion hormone," which is furnished by the medial neurosecretory cells of the brain and released from the corpus cardiacum. At this point, the pupal behavior is turned off and gives way to the appropriate adult pattern.

In analogy with the mammalian evidence for nonconventional interneuronal communication referred to above, it is intriguing that the eclosion hormone, itself a neurochemical mediator, controls this shift in behavior patterns by a direct influence on neural effector cells. It triggers the movements, preprogrammed in the abdominal ganglia, that bring about the terminal molt, and then, by turning off the responsible motor neurons, initiates the degeneration of the abdominal muscles.

Subsequent reproductive events are governed by two additional neurohormonal factors, both released from intrinsic cells of the corpus cardiacum under the direction of the brain. In virgin females, the "calling hormone" brings about the release of a sex pheromone and the assumption of a "calling posture," at a time that coincides with the temperature-dependent flight activity of the male, and thus increases the probability of a rendezvous between the sexes (Truman, 1973b). After mating, which in due course effects the secretion of a blood-borne substance by the bursa copulatrix, the brain dictates a switchover to an "oviposition-stimulating hormone," and this, in turn, is responsible for an increase in the rate of oviposition.

Other impressive examples of behavioral patterns associated with reproductive events occur among blattarians (Barth, 1968). Depending on the species, the male may respond to a volatile pheromone released by the female at the appropriate stage of gonadal maturity and may initiate courtship behavior which includes the dispatch of a male sex pheromone. Control over this ritual and the subsequent copulation and insemination are mediated by hormone action on the part of the corpora allata and not on the part of the gonads.

The receptiveness of the female ceases after the receipt of a spermatophore, and she remains refractory at least until oviposition, or parturition, has taken place. In some forms (*Gromphadorhina*), tactile stimuli prevail, and the sexual dominance of males is linked with territorial and fighting behaviors (Ziegler, 1972). Some grasshoppers engage in courtship and copulation after the male and female have demonstrated their readiness by stridulation (sound production by rubbing of hind legs). This somewhat stereotyped, species-specific behavior pattern likewise depends on the activity of the corpus allatum hormone, which is also responsible for the development of the oocytes. Here too, the females show a sudden shift from responsive to defensive behavior after copulation, brought about by the presence of the spermatophore in the receptaculum seminis, an organ which is innervated by the last abdominal ganglion. Presumably the resulting afferent mechanical stimuli elicit a reversal in the neurally transmitted directives to the corpora allata (Loher & Huber, 1966).

But here, as well as in the blattarians already referred to, the mechanism by which this is accomplished requires further study. Copulatory behavior in mantids, however, seems to be stimulated by a hormone released from the corpus cardiacum.

Crustaceans show a variety of sexual patterns ranging from gonochoristic to hermaphroditic forms (Charniaux-Cotton, 1975). In higher orders, especially decapods, success in the search for a prospective sex partner depends on the readiness of the female, which rejects the male not only before she is ready to spawn but also after copulation has been completed.

Many crustacean males are known to fight with rivals over females. The repertoire of courtship displays may include visual and tactile stimulation, dancing and other postural rituals, stridulation, sometimes transmitted as substrate vibration (Klaassen, 1973), and the use of sex attractants. Parental care consists of various arrangements for carrying, cleaning, and aerating the developing eggs, which may take place in a marsupium, or may involve the building of nests in favorable surroundings (see Schöne, 1961; Bauchau, 1966).

The dependency of these activities on the neuroendocrine apparatus has been demonstrated in various ways—for example, by transplantation experiments leading to masculinization of immature females. Again, two types of endocrine factors, neural and nonneural, are known to participate in this control.

In many of the higher forms, sex determination and the development of male sex characteristics are governed by a hormone originating in the androgenic gland, a structure which is spatially separated from the gonad. The activity of this male sex hormone in very young decapods is restrained by a neurohormone originating in neurosecretory cells of the eyestalk. Similarly, the timetable of ovarian maturation and of subsequent activity cycles is determined by this "gonadinhibiting hormone," the difference being that in females the effect seems to be direct. On the other hand, the presence of the eyestalks is required to insure the development of the external female sex characters. Furthermore, a second neurohormonal factor, derived from the brain and thoracic ganglion, seems to promote ovarian activity in adults (see Adiyodi & Adiyodi, 1970; Démeusy, 1970).

As demonstrated in growing males of gonochoristic amphipods as well as in protandric hermaphrodites during the male phase, the androgenic hormone not only directs the brain to release a neurohormone which stimulates the testicular tissue, but it also effects the liberation of the male gametes. Degeneration of the androgenic gland in hermaphrodites initiates the female phase, during which the absence of neurohormonal stimuli causes the male component of the gonad to regress (Berreur-Bonnenfant, 1967).

Data on comparable control systems in other arthropods are still scanty. An example is the demonstration by Eisen, Warburg, and Galun (1973) of apparent relationships between neurosecretory activity and feeding, mating, and ovipositional behaviors in a representative of the *arachnids*. A further example is the demonstration of an inhibitory action by the neurosecretory system of the chilopod *Lithobius* on the spermatogenetic cycle (Descamps, 1975).

Another group of invertebrates in which a neuroendocrine pattern comparable to those in arthropods has been observed are the *mollusks*. In *Octopus*, and probably also in related cephalopods, the control over sexual maturation is a two-step operation. Release of a gonadotropic hormone by the optic gland, an organ of neural derivation, is programmed by inhibitory neural stimuli (Wells & Wells, 1969). Therefore, in immature animals, severance of the responsible nervous pathways leads to sexual precocity via activation of the optic gland. The dependency of this neural input on photic cues is evident from the fact that operations which blind the animal but leave the gland's innervation intact also bring about gonadal stimulation. No organ comparable to the optic gland has been found in any other group of mollusks.

The "bag cells," a cluster of neurosecretory neurons in the abdominal ganglion of the opisthobranch snail *Aplysia*, phasically release a hormone that is directly responsible not only for the seasonal shedding of eggs but also for the stereotyped behavior pattern accompanying this process (Kupfermann, 1970).

Among protandric hermaphroditic gastropods, such as the slug *Arion*, neurosecretory elements in the brain and tentacles seem to govern sex reversal by exerting differential restraints on the development of the components of the ovotestis (Pelluet, 1964).

Neuroendocrine control of reproduction occurs also among *echinoderms*. In several species of starfishes, a "maturation-inducing substance" of gonadal origin is responsible not only for the shedding of ripe eggs and sperm but also for spawning and brooding behaviors. The activation of this factor during a short specified period is brought about by a neurohormone furnished by the radial nerves (Chaet, 1966; Kanatani, 1973).

In *worms* gametogenesis and related behavioral parameters are under neurohormonal control. Sexual maturation in juvenile nereids (polychetes) is governed by an inhibitory neurosecretory principle, whose concentration gradually decreases.

Either the same hormone, in low concentration, or possibly a separate "maturation hormone," also of cerebral origin, seems to provide the stimulation required by male and female germ cells during a later phase (Durchon, 1967; Hauenschild, 1966, 1974, 1975).

The active principle (or principles) involved seems to be derived, at least in some of the nereids studied, from the numerous neurosecretory cells within the brain rather than the attached "infracerebral gland," the nature of which is still obscure (Baskin, 1970, 1976; Golding, 1974). In dimorphic species, these phenomena are linked with a somatic transformation from the atokous to the epitokous (heteronereid) form and with simultaneous behavioral changes dictated by a shift from the sedentary to a pelagic form of existence.

A most interesting representative of this group of annelids is the palolo worm (*Eunice*), in which swarming of severed body segments and spawning are restricted to a short period of the year timed in relation to the lunar cycle. There is good circumstantial evidence that photoperiodic suppression of neurohormone release is part of the calendar of events responsible for the maturation of the gametes destined to be shed at this time (Hauenschild, Fischer, & Hofmann, 1968). Further gonadotropic effects of neurosecretory mediators have been reported in oligochetes (Herlant–Meewis, 1962; Lattaud, 1974, 1975), hirudineans (Hagadorn, 1969), nemerteans (Bierne, 1970), and planarians (Grasso, 1975; Grasso & Quaglia, 1971; Grasso & Benazzi, 1973).

Even in the most primitive metazoans, the coelenterates, certain evidence suggests the participation of neurosecretory factors in the induction of sexuality and sex differentiation (Burnett & Diehl, 1964).

Conclusion

There is a substantial body of information on the role of hormonal factors in controlling the activities of the central nervous system, including those responsible for behavioral adjustments. New insights have been gained regarding the capacity of the neural apparatus for receiving afferent endocrine signals and for dispatching stimulatory as well as inhibitory commands to the endocrine system. Most of the efferent stimuli involved are mediated by blood-borne neurochemical messengers furnished by specialized (neurosecretory) neurons. Their central role is illustrated by the fact that phylogenetically such neurohormones have preceded the endocrine principles of nonneural origin. In the lower phyla, neurosecretory substances are the only products of internal secretion known to exist, and their influence on reproductive behavior appears to be direct.

In higher animals, such so-called first-order neuroendocrine systems also occur. Apparently, they are carry-overs from the time when endocrine function originated in the nervous system. However, neuroendocrine interactions of increasing complexity and diversity have gained the upper hand and guarantee the governance of sex-related behavior throughout the evolutionary scale. Second-order control systems, linking neural with nonneural endocrine centers, have been demonstrated in higher invertebrates as well as vertebrates, but only in the latter does the participation of sex hormones become an essential and universal feature.

Of particular interest in the present context is the fact that conventional neurons are included among the recipients of neurosecretory signals. Indications for direct effects of such neurochemical factors on behavior exist in insects as well as mammals.

It has been the aim of the present discussion to underscore the significance of neuroendocrine interactions in the governance of reproductive phenomena in invertebrates as well as vertebrates. Even though caution must be exercised in the interpretation of data from such widely divergent groups of animals, the remarkable parallelisms strengthen the concept of central control of individual behavior.

References

Adiyodi, K. G., & Adiyodi, R. G. Endocrine control of reproduction in decapod crustacea. *Biological Reviews*, 1970, *45*, 121–165.

Barry, J., Dubois, M. P., & Poulain, P. LRF producing cells of the mammalian hypothalamus. A fluorescent antibody study. *Zeitschrift für Zellforschung*, 1973, *146*, 351–366.

Barth, R. H. The comparative physiology of reproductive processes in cockroaches. I. Mating behavior and its endocrine control. In A. McLaren (Ed.), *Advances in reproductive physiology*. Vol. 3, 167–207. London: Logos Press, 1968.

Baskin, D. G. Studies on the infracerebral gland of the polychaete annelid, *Nereis limnicola*, in relation to reproduction, salinity, and regeneration. *General and Comparative Endocrinology*, 1970, *15*, 352–360.

Baskin, D. G. Neurosecretion and the endocrinology of nereid polychaetes. *American Zoologist*, 1976, *16*, 107–124.

Bauchau, A. *La vie des crabes*. Paris: Paul Lechevalier, 1966.

Berreur–Bonnenfant, J. Action de la glande androgène et du cerveau sur la gametogenèse de crustacés péricarides. *Archives de Zoologie Expérimentale et Générale*, 1967, *108*, 521–558.

Bierne, J. Recherches sur la différenciation sexuelle au cours de l'ontogenèse et de la régénération chez le némertien *Lineus ruber* (Muller). *Annales des Sciences Naturelles, Zoologie et Biologie Animale*, 1970, *12*, 181–298.

Burnett, A. L., & Diehl, N. A. The nervous system of *Hydra*. III. The initiation of sexuality with special reference to the nervous system. *Journal of Experimental Zoology*, 1964, *157*, 237–249.

Chaet, A. B. Neurochemical control of gamete release in starfish. *Biological Bulletin*, 1966, *130*, 43–58.

Charniaux-Cotton, H. Hermaphroditism and gynandromorphism in malacostracan crustacea. In R. Reinboth (Ed.), *Intersexuality in the animal kingdom*, pp. 91–105. Berlin, Heidelberg, New York: Springer–Verlag, 1975.

Démeusy, N. Quelques aspects de la sexualité chez les décapodes brachyoures gonochoriques. *Bulletin de la Société Zoologique de France*, 1970, *95*, 595–612.

Descamps, M. Etude du contrôle endocrinien du cycle spermatogénétique chez *Lithobius forficatus* L. (Myriapode Chilopode). Rôle du complexe "Cellules neurosécrétrices des lobes frontaux du protocérébron—Glandes cérébrales." *General and Comparative Endocrinology*, 1975, *25*, 346–357.

De Wied, D., van Wimersma Greidanus, T. B., Bohus, B., Urban, I. & Gispen, W. H. Vasopressin and memory consolidation. In M. A. Corner & D. F. Swaab (Eds.), *Progress in brain research*, Vol. 45, *Perspectives in brain research*, pp. 181–194. Amsterdam, London, New York: Elsevier/North Holland Biomedical Press, 1976.

Durchon, M. L'endocrinologie des vers et des mollusques. In *Les grands problèmes de la biologie*, 241 pp. Paris: Masson, 1967.

Ehrensing, R. H., & Kastin, A. J. Melanocyte-stimulating hormone-release inhibiting hormone as an antidepressant: A pilot study. *Archives of General Psychiatry*, 1974, *30*, 63–65.

Eisen, Y., Warburg, M. R., & Galun, R. Neurosecretory activity as related to feeding and oogenesis in the fowl-tick *Argas persicus* (Oken). *General and Comparative Endocrinology*, 1973, *21*, 331–340.

Friedman, E., Friedman, J., & Gershon, S. Dopamine synthesis: Stimulation by a hypothalamic factor. *Science*, 1973, *182*, 831–832.

Golding, D. W. Survey of neuroendocrine phenomena in nonarthropod invertebrates. *Biological Reviews*, 1974, *49*, 161–224.

Grasso, M. Sexuality and neurosecretion in freshwater planarians. In R. Reinboth (Ed.), *Intersexuality in the animal kingdom*, pp. 20–29. Berlin, Heidelberg, New York: Springer-Verlag, 1975.

Grasso, M., & Benazzi, M. Genetic and physiologic control of fissioning and sexuality in planarians. *Journal of Embryology and Experimental Morphology*, 1973, *30*, 317–328.

Grasso, M., & Quaglia, A. Studies on neurosecretion in planarians. III. Neurosecretory fibres near the testes and ovaries of *Polycelis nigra*. *Journal of Submicroscopic Cytology*, 1971, *3*, 171–180.

Hagadorn, I. R. Hormonal control of spermatogenesis in *Hirudo medicinalis*. II. Testicular response to brain removal during the phase of testicular maturity. *General and Comparative Endocrinology*, 1969, *12*, 469–478.

Hauenschild, C. Der hormonale Einfluss des Gehirns auf die sexuelle Entwicklung bei dem Polychaeten *Platynereis dumerilii*. *General and Comparative Endocrinology*, 1966, *6*, 26–73.

Hauenschild, C. Endokrine Beeinflussung der geschlechtlichen Entwicklung einiger Polychaeten. *Fortschritte der Zoologie*, 1974, *22*, 75–92. Stuttgart: Gustav Fischer Verlag.

Hauenschild, C. Die Beteiligung endokriner Mechanismen an der geschlechtlichen Entwicklung und Fortpflanzung von Polychaeten. *Verhandlungen der Deutschen Zoologischen Gesellschaft 1974*, 1975, 292–308.

Hauenschild, C., Fischer, A., & Hofmann, D. K. Untersuchungen am pazifischen Palolowurm *Eunice viridis* (Polychaeta) in Samoa. *Helgolaender Wissenchaftliche Meeresuntersuchungen*, 1968, *18*, 254–295.

Herlant–Meewis, H. Endocrine relationships between nutrition and reproduction in the oligochaete, *Eisenia foetida. General and Comparative Endocrinology*, 1962, *2*, 608.

Horst, W. D., & Spirt, N. A possible mechanism for the anti-depressant activity of thyrotropin releasing hormone. *Life Sciences*, 1974, *15*, 1073–1082.

Kanatani, H. Maturation-inducing substance in starfishes. *International Review of Cytology*, 1973, *35*, 253–298.

Kato, J., Atsumi, Y., & Inaba, M. Estradiol receptors in female rat hypothalamus in the developmental stages and during pubescence. *Endocrinology*, 1974, *94*, 309–317.

Klaassen, F. Stridulation und Kommunikation durch Substratschall bei *Gecarcinus lateralis* (Crustacea Decapoda). *Journal of Comparative Physiology*, 1973, *83*, 73–79.

Kupfermann, I. Stimulation of egg laying by extracts of neuroendocrine cells (bag cells) of abdominal ganglion of *Aplysia. Journal of Neurophysiology*, 1970, *33*, 877–881.

Lattaud, M. Etude en culture organotypique du contrôle du sexe des gamétogenèses chez l'Annelide Oligochète *Eisenia foetida f. typica* Sav.; mise en évidence d'une action androgène des tissus testiculaires en présence du système nerveux central. *Comptes rendus de l'Academie des Sciences, Série D*, 1974, *279*, 935–938.

Lattaud, M. Study of sex control of gametogenesis by organ culture in the oligochaete annelid *Eisenia foetida f. typica* Sav. In R. Reinboth (Ed.), *Intersexuality in the animal kingdom*, pp 64–71. Berlin, Heidelberg, New York: Springer-Verlag, 1975.

Loher, W., & Huber, F. Nervous and endocrine control of sexual behavior in a grasshopper (*Gomphocerus rufus* L., Acridinae). *Symposium of the Society for Experimental Biology*, 1966, *20*, 381–400.

Meyer, C. C. Testosterone concentration in the male chick brain: An autoradiographic survey. *Science*, 1973 *180*, 1381–1383.

Naisse, J. Rôle des neurohormones dans la différenciation sexuelle de *Lampyris noctiluca. Journal of Insect Physiology*, 1969, *15*, 877–892.

Pelluet, D. On the hormonal control of cell differentiation in the ovotestis of slugs (Gastropoda: Pulmonata). *Canadian Journal of Zoology*, 1964, *42*, 195–199.

Pfaff, D. W. Luteinizing hormone-releasing factor potentiates lordosis behavior in hypophysectomized ovariectomized female rats. *Science*, 1973, *182*, 1148–1149.

Plotnikoff, N. P., Kastin, A. J., Anderson, M. S., & Schally, A. V. Oxotremorine antagonism by a hypothalamic hormone, melanocyte-stimulating hormone release-inhibiting factor (MIF). *Proceedings of the Society for Experimental Biology and Medicine*, 1972, *140*, 811–814.

Scharrer, B. Neurohumors and neurohormones: Definitions and terminology. *Journal of Neuro-Visceral Relations, Supplement IX*, 1969, 1–20.

Scharrer, B. General principles of neuroendocrine communication. In F. O. Schmitt (Ed.), *The neurosciences: Second study program*, pp. 519–529. New York: The Rockefeller University Press, 1970.

Scharrer, B. Neuroendocrine communication (neurohormonal, neurohumoral, and intermediate). In J. Ariëns Kappers & J. P. Schadé (Eds.), *Progress in brain research*. Vol. 38, *Topics in Neuroendocrinology*, pp. 7–18. Amsterdam, London, New York: Elsevier, 1972.

Scharrer, B. The concept of neurosecretion past and present. In *Recent studies of hypothalamic function. International Symposium of Calgary 1973,* pp. 1–7. Basel: Karger, 1974a.

Scharrer, B. The spectrum of neuroendocrine communication. In *Recent studies of hypothalamic function. International Symposium of Calgary 1973,* pp. 8–16. Basel: Karger, 1974b.

Scharrer, B. The role of neurons in endocrine regulation: A comparative overview. *Proceedings of the Seventh International Symposium on Comparative Endocrinology,* Tsavo Park, Kenya, 1974. *American Zoologist,* Supplement I, 1975, *15*, 7–11.

Scharrer, B. Neurosecretion—Comparative and evolutionary aspects. In M. A. Corner & D. F. Swaab (Eds.), *Progress in brain research*. Vol. 45, *Perspectives in brain research*, pp. 125–137. Amsterdam, London, New York: Elsevier/North Holland Biomedical Press, 1976.

Scharrer, E. Principles of neuroendocrine integration. In *Endocrines and the central nervous system. Research Publications, Association for Research in Nervous and Mental Disease*, 1966, *43*, 1–35.

Scharrer, E., & Scharrer, B. *Neuroendocrinology*. New York, London: Columbia University Press, 289 pp. 1963.

Schöne, H. Complex behavior. In T. H. Waterman (Ed.), *The physiology of crustacea*. Vol. 2, pp. 465–520. New York, London: Academic Press, 1961.

Truman, J. W. How moths "turn on": A study of the action of hormones on the nervous system. *American Scientist*, 1973a, *61*, 700–706.

Truman, J. W. Temperature sensitive programming of the silkmoth flight clock: A mechanism for adapting to the seasons. *Science*, 1973b, *182*, 727–729.

Truman, J. W., & Riddiford, L. M. Hormonal mechanisms underlying behaviour. In J. E. Treherne, M. J. Berridge, & V. B. Wigglesworth (Eds.), *Advances in insect physiology*. Vol. 10, pp. 297–352. New York, London: Academic Press, 1974.

Weisz, J., & Gibbs, C. Conversion of testosterone and androstenedione to estrogens *in vitro* by the brain of female rats. *Endocrinology*, 1974, *94*, 616–620.

Wells, M. J., & Wells, J. Pituitary analogue in the octopus. *Nature*, 1969, *222*, 293–294.

Ziegler, R. Sexual- und Territorialverhalten der Schabe *Gromphadorhina brunneri* Butler. *Zeitschrift für Tierpsychologie*, 1972, *31*, 531–541.

Zigmond, R. E., Nottebohm, F., & Pfaff, D. W. Androgen-concentrating cells in the midbrain of a songbird. *Science*, 1973, *179*, 1005–1007.

Zimmerman, E. A. Localization of hypothalamic hormones by immunocytochemical techniques. In L. Martini & W. F. Ganong (Eds.), *Frontiers in neuroendocrinology*. Vol. 4, pp. 25–62. New York: Raven Press, 1976.

6

Reproductive Behavior in a Neuroendocrine Perspective

Julian M. Davidson

My objective in this brief presentation is to describe some of the ways in which one can look at reproductive behavior within the framework of an overall neuroendocrine scheme, an approach to which Danny Lehrman made such pioneering contributions. Figure 1 portrays in simplified form various interrelationships among environment, reproductive physiology, and behavior. The diagram shows two feedback loops, one neuroendocrine and the other behavioral, as well as their links to environmental stimuli. According to the view I am presenting, information from the environment is seen as being conveyed, via various sensory modalities, eventually to two separate hypothalamic mechanisms involved in regulating sexual behavior and pituitary gonadotropic function.

The way in which the brain controls adenohypophyseal function is via the release of hypophysiotropic factors into the portal vessel system and thereby to the anterior pituitary. The resulting patterns of circulating gonadotropic hormones control the secretion of gonadal hormones. I shall be concerned here with those arrows in the diagram which suggest actions of gonadal hormones on the two aspects of brain function under discussion: that which links behavior to the brain–gonad loop, and that which represents relationships between the two brain (hypothalamic) mechanisms.

Julian M. Davidson · Department of Physiology, Stanford University, Stanford, California 94305.

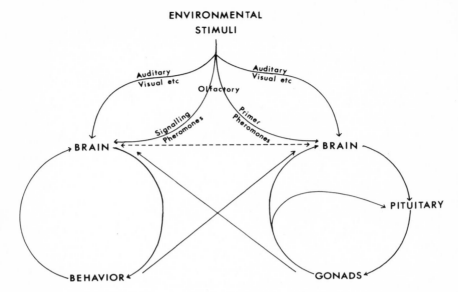

Figure 1. Neuroendocrine and behavioral feedback loops. Environmental stimuli, including "social" stimuli from sexual and nonsexual partners, influence brain mechanisms controlling sexual behavior and the reproductive system, each of which is part of a feedback loop. The two loops are interconnected via the effects of gonadal hormones on sexual behavior and effects of coitus-derived stimuli on neuroendocrine activity. The relations between the two brain systems themselves (depicted here by a broken line) are not yet clear.

Neuroendocrine Feedback for Gonadotropin Regulation

While the importance of feedback mechanisms involving go-nadal hormones and the pituitary are universally recognized, there is no general agreement yet on the extent to which feedback receptors for these responses are located in the brain. It seems likely, in fact, that some of these processes take place in the anterior pituitary it-self—particularly, in our view, positive feedback effects of estrogen (see Smith & Davidson, 1974a).

The conclusion that feedback receptors exist in the hypotha-lamus for the negative feedback regulatory effects of male and female gonadal hormones is based on several lines of evidence. Our own work has concentrated on direct intracranial implantation of the crystalline steroids. The most effective locus for implantation in these cases is the medial basal hypothalamus or median eminence region. However, materials placed in this area could diffuse through the portal vessels to the pituitary. Although implants in the adenohypophysis are generally less effective, this could conceivably be due to relatively poor diffusion of the steroids through pituitary tissue (Bogdanove, 1964). This proposed implantation paradox has to be refuted before the intracranial implanatation experiments can be said to demonstrate the existence of steroid feedback receptors. While in earlier work a variety of indirect methods were used to deal with this question, in recent years the availability of synthetic luteinizing hormone–releasing hormone (LHRH) and of radioimmunoassay has permitted the use of a more direct experimental approach.

If, in fact, the hypothalamically implanted hormone acts by diffusion to the pituitary through its vascular supply, the gland should become insensitive to its own controller, LHRH. We have studied the changing pattern of basal LH released by pituitaries of castrated male and female rats following median eminence implantation of various steroids in crystalline form and simultaneously the responsiveness of the pituitary to LHRH. Testosterone propionate (TP) decreased the LHRH response after a day or more of implantation, but, for a period of up to one day following implantation, we found that the pituitary response was normal (Davidson, Cheung, Smith, & Johnston, 1973a; Cheung and Davidson, 1977). It therefore appeared that the delayed effect of reduced LHRH sensitivity was due to a secondary hypofunction resulting from primary removal of endogenous LHRH by the implant. For the case of ovarian steroids, it was found that implants of estradiol benzoate (EB) or progesterone, although they severely and progressively depressed basal LH secretion, did not chronically inhibit pituitary sensitivity in either male or female rats (Smith & Davidson, 1974a). On the basis of data such as these we conclude that, in fact, the primary action of the steroids in inhibiting gonadotropin secretion (negative feedback) is at the basal hypothalamic level.

The Steroid-Sensitive Mechanism in Male Sexual Behavior

Similar intracerebral implantation experiments have been used to support the existence of putative behavioral receptors for gonadal steroid hormones which are involved in activation of sexual behavior mechanisms. While the hypothalamus is again the important brain area, experiments on male and female rats show that it is the anterior hypothalamic–preoptic region rather than the medial basal hypothalamus which seems to house these steroid-sensitive mechanisms. It is interesting to note that this same region has been implicated in hormonal control of female sexual behavior in cats (Harris & Michael, 1964) as well as of female (Lisk, 1962) and male (Davidson, 1966) sexual behavior in the rat.

Apart from this difference in location, a variety of other interesting differentiations can be made between the feedback and the behavioral mechanisms. Reasonable doses of dihydrotestosterone (DHT), a potent androgen which may be the active intermediate product for testosterone's action on peripheral sexual tissues, do not maintain copulatory behavior in the male rat (Feder, 1971) and are relatively ineffective relating to the sexual behavior of various other species (references in Alsum & Goy, 1974).* DHT is, however, a potent feedback-active agent in the control of gonadotropin secretion in all cases which have been studied, as shown by injection or hypothalamic implantation experiments (Johnston & Davidson, 1972; Smith & Davidson, 1974a). The low behavioral effectiveness of this 5α-reduced androgen is not due to its failure to reach the brain. In fact, direct intrahypothalamic implantation of dihydrotestosterone has very little activating effect on copulatory behavior in castrate male rats when compared to testosterone (Johnston & Davidson, 1972).

It has now been repeatedly demonstrated that male sexual behavior can be effectively activated in castrate rats by combining a small dose of estrogen with injections of DHT (Feder, Naftolin, & Ryan, 1974; Baum & Vreeburg, 1973; Larsson, Södersten, & Beyer, 1973). This has been used to support the concept that androgenic action on the brain requires aromatization to estrogen, while its action on the genitals is via 5α-reduction to the dihydro form. Thus the activation of behavior by the combined action of the two steroids could be due to simultaneous stimulation of brain by estrogen and of periphery (penis) by DHT. However, we feel that this hypothesis is not yet established, and recent data suggest that other formulations relating to

* One exception is the guinea pig (Alsum & Goy, 1974). DHT was also found to be effective in one and ineffective in another mouse strain (Luttge & Hall, 1973), and was about half as potent as T in the rhesus (Phoenix, 1974).

the site(s) of steroid action on male sex behavior may be preferable.

First, in a recent study Baum, Södersten, and Vreeburg (1974) suggest that the two steroids may act in synergistic fashion at the central level, as indicated by the failure of genital anesthetization to eliminate the effect of the combined treatment. Second, we administered DHT to castrate male rats with intrahypothalamic TP implants and obtained no significant improvement in the (suboptimal) copulatory behavior found in these animals (Davidson & Trupin, 1975). While this does not disprove that peripheral androgen action has a role in male sex behavior, it at least shows that the behavioral deficit in animals with anterior hypothalamic implants of testosterone is not due to failure of peripheral (i.e., penile) mechanisms. There is still no sufficient reason to believe that concurrent (as opposed to perinatal) stimulation of the extracerebral tissues by androgen in the adult is of any major import for male sexual behavior in the rat, despite the favored interpretation of the results of DHT–estrogen experiments suggested in recent reports.

Finally, we have treated prepuberally castrated adult male rats with subcutaneous Silastic implants of DHT combined with low doses of EB injected (0.5μg/day) or implanted in crystalline form in the anterior hypothalamic region (Davidson & Trupin, 1975). The combined intracerebral-EB–peripheral-DHT treatment did activate male copulatory responses to an extent similar to that found with intrahypothalamic TP alone. Maximal behavioral responsiveness was reached more slowly but probably was maintained longer than with TP implantation. Strangely enough, when the DHT capsules were removed, mating behavior did not deteriorate for at least 4 weeks, during which time the frequency of mounts, intromissions, and ejaculations remained unchanged. This indicates that concurrent stimulation of the periphery by DHT is unnecessary following prior exposure to combined EB–DHT. It seems to us that the most reasonable interpretation is that DHT, rather than acting purely peripherally in conjunction with the central action of EB, synergizes with it at the level of the CNS, as suggested by Baum *et al.* (1974) (see above).

In addition to studying their different responses to natural steroids, one can compare the feedback and behavioral systems in the male by studying the relative effectiveness of antiandrogens on these two types of responses. We have previously studied cyproterone, a powerful steroidal androgen antagonist which has no antiandrogenic properties on sexual behavior in the rat but which does activate the feedback mechanism for gonadotropin secretion by inhibiting the suppressive effects of endogenous androgen (von Berswordt-Wallrabe & Neumann, 1967; Bloch & Davidson, 1967). We wondered whether this differential effect of the antiandrogen on the two systems was a general property of their responses to androgen antagonists rather than an idiosyncracy of the cyproterone molecule. Accordingly, we welcomed the opportunity to work with a new antiandrogen of rather different (nonsteroidal) molecular structure, Flutamide (4'-nitro-3'-trifluoromethylisobutyranilide). The results were clear-cut, as far as feedback is concerned: a very large increase in both LH and testosterone secretion was found following chronic injection of 50 mg/kg/day into adult male rats. However, when this large dose was administered for 28 days to experienced adults, there was no inhibition of behavioral responses, but rather a slight enhancement, which may have been an effect of experience. These experiments and other differences between the behavioral and feedback systems are discussed in Davidson and Trupin (1975).*

Multiple Roles of Behavior in Neuroendocrine Systems

Referring once again to Figure 1, we note the arrow which depicts the other direction of the hormone–behavior relationship: effects of behavioral stimuli on hormone secretion. The particular situation which I would like to describe is that of the effects of copulatory stimuli on gonadotropin secretion in female rats in which spontaneous ovulation is blocked, specifically by constant illumination. This work was done in Oxford with K. Brown-Grant and in Stanford with E. R. Smith.

Similar to the observations of Tom McGill on luteal activation in the mouse (mentioned earlier in this conference), here too intromissions are the most effective stimulus for ovulation and LH release. Ejaculation is unnecessary, and mounts alone are only partially effec-

---

*Since the conference, we have studied the behavioral effects of the same dose of Flutamide in long-term castrates treated with a dose of testosterone propionate just adequate to restore normal copulatory behavior in 100% of castrated rats. Although ejaculatory responses were significantly inhibited, intromission and mounting behavior was unchanged though seminal vesicle, prostate, and penis weights were indistinguishable from those of untreated castrates (Södersten, Gray, Damassa, Smith, & Davidson, 1975).

tive (Brown-Grant, Davidson, & Greig, 1973). The dynamics of hormone release in this situation are such that, 10 minutes following the onset of mating, plasma LH shows a significant increase and the peak occurs in about 1 hour. The effect on FSH is very small (Brown-Grant *et al.*, 1973) or absent (Davidson, Smith, & Bowers, 1973b), but prolactin shows a rapid rise, peaking within 10 minutes in our constant-light-exposed rats. Prolactin is an extremely stress-responsive hormone, and much of the apparent effect of mating seems to have been due to nonspecific stimulation (Davidson *et al.*, 1973b). We have also studied LHRH concentration in the hypothalamus following mating. There was an initial depletion, presumably due to rapid release of stored peptide, followed by evidence of resynthesis (Smith & Davidson, 1974b).

We have studied another rat preparation in which one can demonstrate mating-induced LH release: Ovariectomized rats treated with high doses of estrogen to make them sexually receptive and suppress LH secretion will also show an LH surge following mating (Davidson *et al.*, 1973a; Davidson & Smith, 1974). Interestingly, there is an interaction between mating-induced reflex LH release and the diurnal factor, which tends under various circumstances to facilitate LH release in the latter part of the light period (Caligaris, Astrada, & Taleisnik, 1971). Thus we found that mating-induced LH release in these spayed, estrogen-treated rats was much higher in the afternoon than in the morning (Davidson & Smith, 1974).

Finally, I should like to indicate some of the complexities of behavior–endocrine interactions by an example from the primate literature. Androgen levels are known to be related causally to aggressive behavior in some species. In rodents, for instance, castration and replacement therapy with androgen respectively decreases or restores agonistic behavior (see Hart, 1974). Work on rhesus monkeys from the laboratory of Rose shows that the situation, in primates at least, is far from that of a simple causal relationship. Adult male monkeys placed with a group of females (during which time they copulated frequently) showed sharp increases in testosterone level and a return to baseline after removal from the female group (Rose, Gordon, & Bernstein, 1972). This endocrine response could have been the direct consequence of coitus, as has been demonstrated for bulls (Katongole, Naftolin, & Short, 1971), rabbits (Endroczi & Lissak, 1962), and rats (Purvis & Haynes, 1974). However, the male monkeys assumed a socially dominant position over the females. Was it coitus per se or other social stimuli which triggered the change in testosterone?

Subsequent studies (Rose, Bernstein, & Gordon, 1975) have provided evidence that conflict involving alteration in status of macaques living in social groups can produce large swings in testosterone level, regardless of sexual stimuli. Increases resulted when the animal became dominant and decreases occurred in the case of submission. The drop in testosterone concentration following defeat might well be related to the phenomenon demonstrated in male humans (Carstensen, Amer, Wide, & Amer, 1973) and rats (Bliss, Frischat, & Samuels, 1972) of a drop in blood testosterone following various surgical, psychogenic, or other stresses. As suggested by Rose *et al.* (1975), a type of positive feedback may be operating in this situation. Thus, defeat-induced decreases in hormone level may reduce the likelihood that the submissive male will show further aggressive behavior (by direct action on the brain), thus reducing the likelihood of further damaging encounters. Similarly, one might add that, if indeed heterosexual coitus raises testosterone level chronically (Drori & Folman, 1964), this in turn could increase the probability of further copulation, thus maintaining a high level of sexual activity in successful males living with a group of females.

Thus the testosterone level can function in these situations as a dependent and/or independent variable. One has to realize, however, that any discussion of mechanisms for these phenomena is in the realm of pure speculation. The stress-induced reduction in blood testosterone, for instance, is a most obscure phenomenon. It apparently does not depend on decreased gonadotropin secretion [Carstensen *et al.* (1973) and unpublished data from our laboratory].* If it is a nonspecific effect of stress, it is not clear why winning a fight should have the opposite effect to losing. Social-dominance-related rises in testosterone unconnected with sexual stimuli remain to be substantiated. We must also consider that the situation may be quite different in stable hierarchies, where fighting may be minimal, than in changing social situations. In a recent investigation, Eaton and Resko (1974) could find no correlation between established dominance rank

---

* Since the conference, we have shown LH decreases accompanying the testosterone suppression.

and blood testosterone (one determination only per animal) in a captive natural troop of Japanese macaques.

Clearly, the precise elucidation of the operation of complex sociobehavioral neuroendocrine systems such as this will involve a great deal of careful work during periods of social change and through the establishment of steady-state situations. This type of endeavor can certainly provide employment to many behavioral endocrinologists for the foreseeable future.

ACKNOWLEDGMENTS

This manuscript, supported by USPHS grants MH21178 and HD778, was prepared during the author's tenure as a visiting fellow at Battelle Seattle Research Center.

References

Alsum, P., & Goy, R. W. Actions of esters of testosterone, dihydrotestosterone or estradiol on sexual behavior in castrated male guinea pigs. *Hormones and Behavior*, 1974, *5*, 207–217.

Baum, M. J., & Vreeburg, J. T. M. Copulation in castrated male rats following combined treatment with estradiol and dihydrotestosterone. *Science*, 1973, *182*, 283–284.

Baum, M. J., Södersten, P., & Vreeburg, J. T. M. Mounting and receptive behavior in the ovariectomized female rat: Influence of estradiol, dihydrotestosterone and genital anesthetization. *Hormones and Behavior* 1974, *5*, 175–190.

Bermant, G., & Davidson, J. M. *Biological bases of sexual behavior.* New York: Harper and Row, 1974.

Berswordt-Wallrabe, R. von, & Neumann, F. Influence of a testosterone antagonist (cyproterone) on pituitary and serum FSH-content in juvenile male rats. *Neuroendocrinology* 1967, *2*, 107–112.

Bliss, E. L., Frischat, A., & Samuels, L. Brain and testicular function. *Life Sciences* 1972, *2*, 231–238.

Bloch, G. J., & Davidson, J. M. Antiandrogen implanted in brain stimulates male reproductive system. *Science* 1967, *155*, 593–595.

Bogdanove, E. M. The role of the brain in the regulation of pituitary gonadotropin secretion. *Vitamins and Hormones (New York)*, 1964, *22*, 205–260.

Brown-Grant, K., Davidson, J. M., & Greig, F. Induced ovulation in albino rats exposed to constant light. *Journal of Endocrinology*, 1973, *57*, 7–22.

Caligaris, L., Astrada, J. J., & Taleisnik, S. Biphasic effect of progesterone on the release of gonadotropin in rats. *Endocrinology*, 1971, *89*, 331–337.

Carstensen, H., Amer, I., Wide, L., & Amer, B. Plasma testosterone, LH and FSH during the first 24 hours after surgical operations. *Journal of Steroid Biochemistry*, 1973, *4*, 605–611.

Cheung, C. Y., & Davidson, J. M. Effects of testosterone implants and hypothalamic lesions of luteinizing hormone regulation in the castrated male rat. *Endocrinology*, 1977, *100*(2), 292–302.

Davidson, J. M. Activation of the male rat's sexual behavior by intracerebral implantation of androgen. *Endocrinology*, 1966, *79*, 783–794.

Davidson, J. M., & Cheung, C. Inhibitory feedback action of gonadal steroids at the hypothalamic level. *Federation Proceedings, Federation of American Societies for Experimental Biology*, 1973, *32*, 227 (abstract).

Davidson, J. M., & Smith, E. R. Gonadotropin release as a function of mating and steroid feedback in the female rat. *Hormones and Behavior*, 1974, *5*, 163–174.

Davidson, J. M., and Trupin, S. Neural mediation of steroid-induced sexual behavior in rats. In M. Sandler and G. L. Gessa (Eds.), *Sexual behavior: Pharmacology and biochemistry*. New York: Raven Press, 1975. Pp. 13–20.

Davidson, J. M., Cheung, C., Smith, E. R., & Johnston, P. Feedback regulation of gonadotropins in the male. *Advances in Bioscience*, 1973a, *10*, 63–72.

Davidson, J. M., Smith, E. R., & Bowers, C. Y. Effects of mating on gonadotropin release in the female rat. *Endocrinology*, 1973b, *93*(5), 1185–1192.

Drori, D., & Folman, Y. Effects of cohabitation on the reproductive system, kidneys and body composition of male rats. *Journal of Reproduction and Fertilization*, 1964, *8*, 351–359.

Eaton, G. G., & Resko, J. A. Plasma testosterone and male dominance in a Japanese macaque (*Macacca fuscata*) troop compared with repeated measures of testosterone in laboratory males. *Hormones and Behavior*, 1974, 251–259.

Endröczi, E., & Lissak, K. Role of reflexogenic factors in testicular hormone secretion. Effect of copulation on the testicular hormone production in the rabbit. *Acta Physiologica Academiae Scientiarum Hungaricae*, 1962, *21*, 203–206.

Feder, H. H. The comparative actions of testosterone propionate and 5α-androstan-17β-3-one propionate on the reproductive behavior, physiology and morphology of male rats. *Journal Endocrinology*, 1971, *51*, 241–252.

Feder, H. H., Naftolin, F., & Ryan, K. J. Male and female sexual responses in male rats given estradiol benzoate and 5α-androstan-17β-ol-3-one propionate. *Endocrinology*, 1974, *94*, 136–141.

Harris, G. W., & Michael, R. P. The activation of sexual behavior by hypothalamic implants of oestrogen. *Journal of Physiology*, 1964, *171*, 275–301.

Hart, B. L. Gonadal androgen and sociosexual behavior of male mammals: A comparative analysis. *Psychological Bulletin*, 1974, *81*, 383–400.

Johnston, P., & Davidson, J. M. Intracerebral androgens and sexual behavior in the male rat. *Hormones and Behavior*, 1972, *3*, 345–357.

Katongole, C. B., Naftolin, F., & Short, R. V. Relationship between blood levels of luteinizing hormone and testosterone in bulls and the effects of sexual stimulation. *Journal of Endocrinology*, 1971, *50*, 457–466.

Larsson, K., Södersten, P., & Beyer, C. Sexual behavior in male rats treated with estrogen in combination with dihydrotestosterone. *Hormones and Behavior*, 1973, *4*, 289–299.

Lisk, R. D. Diencephalic placement of estradiol and sexual receptivity in the female rat. *American Journal of Physiology*, 1962, *203*, 493–496.

Luttge, W. G., & Hall, N. R. Differential effectiveness of testosterone and its metabolites in the induction of male sexual behavior in two strains of albino mice. *Hormones and Behavior*, 1973, *4*, 31–43.

Phoenix, C. H. Effects of dihydrotestosterone on sexual behavior of castrated male rhesus monkeys. *Physiology and Behavior*, 1974, *12*, 1045–1055.

Purvis, K., & Haynes, N. B. Short-term effects of copulation, human chorionic gonadotrophin injection and non-tactile association with a female on testosterone levels in the male rat. *Journal of Endocrinology*, 1974, *60*, 429–439.

Rose, R. M., Gordon, T. P., & Bernstein, I. S. Plasma testosterone levels in the male rhesus: Influences of sexual and social stimuli. *Science*, 1972, *178*, 643–645.

Rose, R. M., Bernstein, I. S., & Gordon, T. P. Consequences of social conflict on plasma testosterone levels in rhesus monkeys. *Psychosomatic Medicine*, 1975, *37*, 50–61.

Smith, E. R., & Davidson, J. M. Location of feedback receptors: Effects on plasma LH and LRF response of intracranially implanted steroids. *Endocrinology*, 1974a, *95*, 1566–1573.

Smith, E. R., & Davidson, J. M. Luteinizing hormone releasing factor in constant light exposed rats: Effects of mating. *Neuroendocrinology*, 1974b, *14*, 129–138.

Södersten, P., Gray, G., Damassa, D. A., Smith, E. R., & Davidson, J. M. Effects of a non-steroidal antiandrogen on sexual behavior and pituitary-gonadal function in the male rat. *Endocrinology*, 1975, *97*, 1468–1475.

7

An Evolutionary Approach to Brain Research on Prosematic (Nonverbal) Behavior

Paul D. MacLean

In this paper I deal with the question of forebrain mechanisms underlying species-typical communicative behavior that is basic for self-preservation and the survival of the species. In my own research on this problem I have taken an evolutionary approach, which has the advantage that it allows one to telescope millions of years into a span that can be seen all at once and, as in plotting a curve, to detect trends that would not otherwise be apparent.

Communicative Behavior

Human communicative behavior can be classified as verbal and nonverbal. Like P. W. Bridgman, the physicist–philosopher, people usually assume that "most human communication is verbal" (1959). Contrary to the popular view, many behavioral scientists would place a greater emphasis on nonverbal communication. For example, when four such scientists were asked to draw two squares representative of

Paul D. MacLean · Laboratory of Brain Evolution and Behavior, National Institute of Mental Health, Bethesda, Maryland 20014.

the weight that they would give to verbal and nonverbal behavior in our day-to-day activities, there was a striking similarity in their responses: in each case, the square for nonverbal behavior was about three times greater than the one for verbal behavior.

Many forms of human nonverbal behavior show a close parallel to behavioral patterns seen in animals. Since it is hardly appropriate to refer to nonverbal behavior of animals, we need some other term for this kind of behavior. The Greek word σημη pertains to a sign, mark, or token. By adding the prefix προ in its particular sense of *rudimentary*, one derives the word *prosematic*, which would be appropriate for referring to basic forms of communication involving any kind of nonverbal signal—vocal, bodily, or chemical (MacLean, 1974a, 1975b).

In a very real sense, prosematic behavior, like verbal behavior, can be considered in terms of semantics and syntax. Somewhat analogous to words, sentences, and paragraphs, prosematic behavior becomes meaningful in terms of its components, constructs, and sequences of constructs.

Components of behavior are the small building blocks of behavior, including discrete movements, such as the wink of an eye, or elements of an autonomic response such as the blush caused by the dilation of a single blood vessel. *Constructs* are specific patterns of behavior made up of behavioral components. In generating computer language, the programmer thinks nothing of developing a whole new vocabulary. In like vein, when using ordinary language, there need be no reticence to create words for which adequate terms do not exist. By using the stem for the word *ethology*, we can devise four words to identify four principle constructs that have a signaling value in prosematic communication. The word *ethen* would refer to an autonomic response, such as piloerection or a generalized blush, that has signaling value. *Ethones* would apply to unconditioned responses, such as sneezing and startle, evoked by activation of a special sensory system. *Ethid* would apply to combinations of somatic and autonomic responses expressive of basic bodily needs, including breathing, ingestion, the emunctories, copulation, and the posture of sleep. Finally, *ethons* would refer to species-typical acts serving symbolically as social signals, such as the head-bobbing in a lizard's display or the showing of the neck by a wolf in a standoff situation.

These four main types of constructs, in turn, may be ordered in various sequences or *ethexons*—a term representing a contraction of the Greek words for character and sequence. The ethexons themselves fall into six main categories that may be labeled (1) *searching*, (2) *assertive*, (3) *protective*, (4) *attachment*, (5) *dejected*, and (6) *gratulant* (MacLean, 1970). Since most terrestial vertebrates engage in these forms of behavior, it is not meaningful for our present purpose to refer to them in the traditional manner as *species-specific behavior*. But since various species perform these behaviors in their own typical ways, it is both correct and useful to refer to species-typical behavior. As ethologists have emphasized, a species can be identified as readily by its behavioral patterns as by its morphological characters.

Evolutionary Considerations

The relevance of work on animals to human affairs becomes apparent when it is realized that the primate brain derives from three evolutionary developments characterized as reptilian, paleomammalian, and neomammalian (MacLean, 1958, 1962). The illustration in Figure 1 suggests the hierarchic organization of the three evolutionary formations. Radically different in their chemistry and structure and, in an evolutionary sense, countless generations apart, the three basic formations constitute, so to speak, three brains in one, or what may be appropriately referred to as a *triune* brain (MacLean, 1970, 1973c).

What this situation implies is that we are obliged to look at ourselves and the world through the eyes of three quite different mentalities. Stated in present-day terms, one might imagine that our brain represents an amalgamation of three biological computers, each with its own special kind of intelligence, its own special form of subjectivity, its own sense of time and space, its own type of memory, etc.

The proposed scheme for subdividing the brain may seem simplistic, but, thanks to improved neuroanatomical, histochemical, and physiological techniques, the three basic formations of the forebrain stand out in clearer detail than ever before. Moreover, it should be emphasized that, despite their extensive interconnections, each brain type is capable of operating somewhat independently. Most important, in regard to the verbal–nonverbal question, there are clinical in-

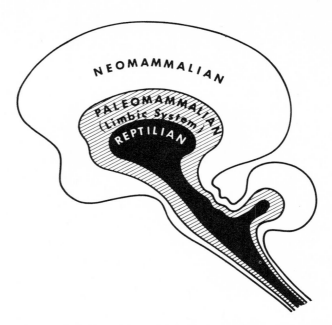

Figure 1. The primate forebrain. In evolution, the primate forebrain expands in hierarchic fashion along the lines of three basic patterns that may be characterized as reptilian, paleomammalian, and neomammalian (from MacLean, 1967).

dications that the reptilian and paleomammalian formations lack the neural machinery for verbal communication. To say that they lack the power of speech, however, does not belittle their intelligence, nor does it relegate them subjectively to the realm of the unconscious.

From an evolutionary standpoint the reptilian brain is of particular interest because it allows one to visualize

> how developments at a critical locus in the so-called hypopallium [See a, Figure 2] described by Elliot Smith (1918–19) may have tipped the scales so that some animals evolved in the direction of birds, while others went the mammalian way. The critical area lies near the ventrolateral base of what J. B. Johnston in 1916 called the dorsal ventricular ridge, presumably because it reminded him of a mountain ridge. In an extension of Johnston's geological analogy, the proliferating hypopallial area might be imagined as comparable to a turbulent volcanic zone. In birds its continued eruption resulted in a piling up of ganglia upon ganglia, whereas its explosion in mammals was responsible for the mushrooming neocortex forming the dorsolateral part of the brain. (MacLean, 1975c)

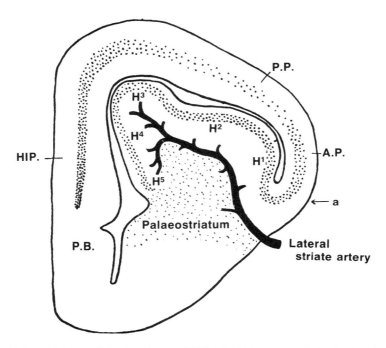

Figure 2. Reproduction of the first figure of Elliot Smith's paper (1918–19) showing a section through the forebrain of tuatara (*Sphenodon punctatum*). See text regarding significance of U-shaped collection of cells indicated by arrow (a). H¹–H⁵ are areas which Elliot Smith included under the designation *hypopallium* and which are part of what J. B. Johnston (1916) had referred to previously as the *dorsal ventricular ridge*. Other abbreviations: A.P., area pyriformis [sic]; HIP., hippocampal formation; P.B., paraterminal body; P.P., parahippocampal pallium.

Ventromedial to the hypopallial zone is a striate mass of structures that has remained firmly embedded in the brains of reptiles, birds, and mammals (see below). In the rest of this paper I am going to focus mainly on this archaic part of the brain. In mammals, the counterpart of the reptilian mass is represented by the olfactostriatum, corpus striatum (caudate nucleus and putamen), globus pallidus, and satellite collections of gray matter. Since there is no term that applies to all of these structures, I will refer to them as the striatal complex.

The black areas in Figure 3 show in the squirrel monkey how a stain for cholinesterase sharply demarcates the striatal complex from

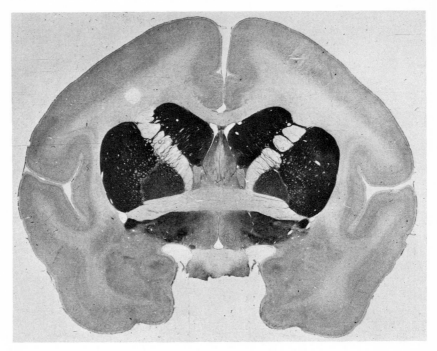

Figure 3. Section from brain of squirrel monkey, illustrating how a stain for cholin-esterase selectively colors (black areas) the striatal complex (from MacLean, 1972a).

the two other evolutionary formations. One of the contributions of histochemistry has been the revelation that the corresponding struc-tures can be similarly identified in all animals ranging from reptiles to man (see Figure 4). With the fluorescent technique of Falck and Hillarp (1959), it is striking to see how the same structures shown in Figure 4 glow a bright green because of the presence of dopamine (Juorio & Vogt, 1967), a neural sap that seems to be of prime impor-tance in the function of brain mechanisms that bring into play the concerted energies of the organism.

Prototypical Behavioral Patterns

From an evolutionary standpoint it is curious that ethologists have paid little attention to reptiles, focusing instead on fishes and birds.* Some authorities believe that, of existing reptiles, lizards

*The wording of this and the next paragraph follows closely that of MacLean (1975b).

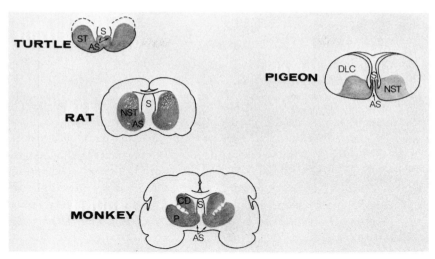

Figure 4. Shaded areas show how a stain for cholinesterase demarcates corresponding structures of the striatal complex in animals ranging from reptiles to primates: With the technique of Falck and Hillarp, the same areas shown above develop a bright green fluorescence because of large amounts of dopamine (adapted from Parent & Olivier, 1970; see MacLean, 1973b, for colored plate).

would bear the closest resemblance to the mammal-like reptiles believed to be the forerunners of mammals. At all events, lizards and other reptiles provide illustrations of complex prototypical patterns of behavior commonly seen in mammals, including man. It is easy to list more than 20 such behaviors that may primarily involve self-preservation or the survival of the species (MacLean, 1975a,b): (1) selection and preparation of homesite; (2) establishment of domain or territory; (3) patrolling territory; (4) trail making; (5) marking of domain or territory; (6) showing place preferences; (7) ritualistic display in defense of territory, commonly involving the use of coloration and adornments; (8) formalized intraspecific fighting in defense of territory; (9) triumphal display in successful defense; (10) assumption of distinctive postures and coloration in signaling surrender; (11) routinization of daily activities; (12) foraging; (13) hunting; (14) homing; (15) hoarding; (16) use of defecation posts; (17) formation of social groups; (18) establishment of social hierarchy by ritualistic display and other means; (19) greeting; (20) grooming; (21) courtship, with displays using coloration and adornments; (22) mating;

(23) breeding and, in isolated instances, attending offspring; (24) flocking; and (25) migration.

Five Interoperative Behaviors

There is an important pentad of prototypical forms of behavior of a general nature that may be variously operative in the above activities. In anticipation of some later comments, I will name and briefly characterize them. They may be denoted as (1) isopraxic, (2) perseverative, (3) re-enactment, (4) tropistic, and (5) deceptive behavior. The word *isopraxic* refers to behavior in which two or more individuals engage in the same kind of activity. As a purely descriptive term, isopraxic avoids preconceptions and prejudices commonly attached to such terms as *social facilitation* and *imitation*. Perseverative behavior applies to repetitious acts like those that occur in displays and would include so-called displacement or adjunctive behavior seen in conflictive situations. Re-enactment behavior refers to the repetition on different occasions of behaviors seeming to represent obeisance to precedent, as, for example, following familiar trails or returning year after year to the same breeding grounds. Tropistic behavior is characterized by positive or negative responses to partial or complete representations, whether alive or inanimate, and includes what ethologists refer to as *imprinting* and *fixed action patterns*. Deceptive behavior involves the use of artifice and deceitful tactics such as are employed in stalking a prey or evading a predator. It is remarkable how many patterns of behavior seen in reptiles are also found in human beings.

Experimental Findings

As yet, hardly any investigations have been conducted on reptiles in an attempt to identify specific structures of the forebrain involved in the various behaviors listed above. All that is known thus far is that the neural guiding systems for complex species-typical behavior lie forward of the midbrain. With Greenberg and Ferguson, I have performed some pilot experiments on the effects of lesions of the striatal complex on the display behavior of the green Anolis lizard (Greenberg, Ferguson, & MacLean, 1976). The results have been of particular interest in regard to the so-called challenge or territorial display characterized by pushups, extension of the throat fan, and profile changes that increase the apparent size of the lizard. Because the optic nerves are almost entirely crossed, we can injure the striatal complex in one hemisphere and then test the animal's display with either eye covered. Although capable of seeing, the animal shows no interest in a rival lizard when looking with the eye projecting to the injured hemisphere. But allow him to see his rival with the intact hemisphere and the challenge display behavior returns immediately in full force.

In contrast to reptiles, the striatal complex of mammals has been subjected to extensive investigation. It cannot be overemphasized, however, that 150 years of experimentation have revealed little specific information about its functions. The finding that large bilateral destructions of the striatal complex may result in no impairment of movement speaks against the traditional clinical view that it subserves purely motor functions. As with reptiles, we are conducting experiments on mammals, testing the hypothesis that the striatal complex plays a basic role in species-typical, prosematic behavior.

Findings in Monkeys

Thus far crucial findings have turned up in the work on squirrel monkeys. It is a remarkable parallel that, as in some reptiles and lower forms, squirrel monkeys utilize a similar display in both courtship and aggression. In each situation (see Figure 5) the male vocalizes, spreads one thigh, and directs the erect phallus towards the other animal (Ploog & MacLean, 1963). The aggressive display is seen in its most dramatic form when a new male is introduced into an established colony of monkeys. Within seconds all the males follow suit in displaying to the strange monkey; if it does not remain quiet with its head bowed, it will be viciously attacked. Ploog and I found that the incidence of the display among males in a colony is a better measure of dominance than the outcome of rivalry for food. Although females are capable of a modified display, they do not join in with the males in displaying to the strange monkey.

The display is also used as a form of greeting. I have described one variety of squirrel monkey that will regularly perform a greeting display upon seeing its reflection in a mirror (MacLean, 1964). We

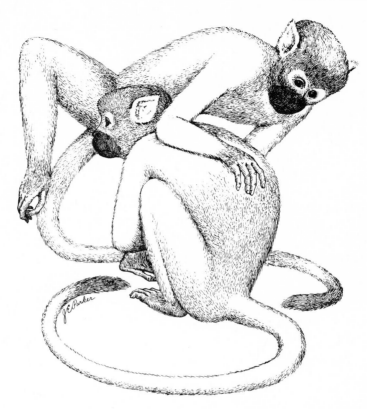

Figure 5. Depiction of aggressive display of a squirrel monkey to a lower-ranking male. A similar display is used prior to attempts at copulation with a female (from Ploog & MacLean, 1963).

refer to the displaying animal shown on the left in Figure 6 as the Gothic type because the ocular patch forms a peak over the eye like a Gothic arch, whereas we call the others Romans because the ocular patch is round like a Roman arch. The display of both varieties is quite similar in the communal situation, but only the Gothic type will regularly display to its reflection in a mirror.

Ploog, Kopf, and Winter (1967) have since observed that, without exposure to any animal other than its own mother, an infant squirrel monkey will display to another monkey as early as the second day of life. This finding clearly indicates that the display is a genetically constituted form of behavior.

Figure 6. Two varieties of squirrel monkeys referred to as "Gothic" (left) and "Roman" (right) because of the pointed and rounded shape of the ocular patch above the eye. Both varieties use the same type of display in the communal situation, but only the Gothic type will consistently display to its reflection in a mirror (from MacLean, 1964).

We have found that, if Gothic monkeys are kept visually isolated, about 80% of them will consistently display to their reflections in a mirror (see Figure 7). Consequently, I have used this test as a model for investigating what parts of the brain may be involved in territorial and courtship display and in greeting. The mirror display test has the advantage that it eliminates olfactory and other cues.

All of the display testing is done in the monkey's home cage with a mechanical contrivance for exposing a full length mirror for 30 seconds. The animals are tested twice a day. Figure 8 shows a typical protocol chart recording the latency and magnitude of the erection, as well as the other main components of the display—namely, vocalization, spreading of the thigh, scratching, and urination. After a monkey reaches criterion of near perfect performance in 30 or more trials, it is subjected to bilateral lesions of target structures and then retested for several weeks or months. These experiments have involved over 30 monkey-years of observation.

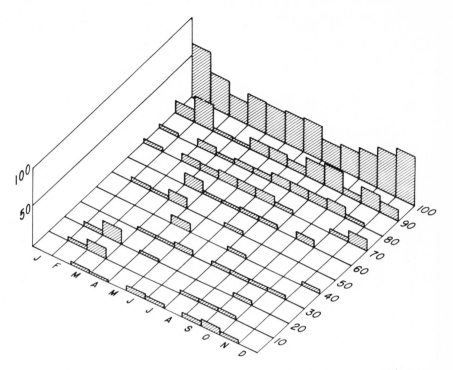

Figure 7. Array of histograms showing the results of mirror testing of unoperated Gothic-type squirrel monkeys during different months of the year. Data were derived from observations on 105 monkeys. Each animal was given a minimum of 30 trials. Criterion was performance of the display in 80% or more of the trials. About 80% of animals achieve this level of performance. The ordinate scale gives percentage values for the number of monkeys tested, and the other numerical scale gives percentage values for scores. Initials identify months of the year. Since the results indicate that a statistically significant number of randomly chosen animals are less likely to display in March, seasonal factors must be taken into consideration in mirror testing.

Using the mirror test, I have observed the effects of various brain lesions on the display of more than 100 monkeys. Large bilateral lesions in the paleo- and neomammalian formations of the brain may have either no effect or only a transitory effect on the display. The same has proved true of complete removal of the superior colliculus and pretectal structures. Figure 9 shows large bilateral lesions of the amygdala in an animal that continued to display consistently day after day just as before operation.

PRE-OP TESTING
TO MECH. MIRROR

#762
Z-THREE

Trial #	Date 1969	Time	Pos	Er	Lat	Ur	Voc	TS	Scr	Cue	Time exp		Comments
1	10/16	11:07	↑	5+	7"	+	+	+	O	E	30"		RG
2	10/19	9:52	↑	5+	8"	⊕	+	+	O	E	30'		RG
3	10/19	1:45	↑	5+	5"	+	+	+	O	E	30"	PULLED ON PENIS	RG
4	10/20	10:11	↑	5+	10"	+	+	+	o	E	50ᵏ		RG
5	10/20	2:04	↑	5+	9'	+	+	+	O	E	30'		RG
6	10/21	11:07	↑	5+	5"	O	+	+	O	E	30'		RG
7	10/21	5:03	↑	5+	5"	+	+	+	O	E	30"		RG
8	10/22	11:25	↑	5+	5'	⊕	+	+	O	E	30ᵏ		RG
9	10/22	2:22	↑	5+	4"	+	+	+	O	E	30ᵏ		RG
10	10/23	12:44	↑	5+	5"	+	+	+	O	E	30'		RG
11	10/23	2:20	↑	5+	5"	+	+	+	O	E	30"		RG
12	10/23	4:43	↑	5+	5"	+	+	+	O	E	30ᵏ		RG
13	10/24	11:44	↑	5+	6"	O	+	+	O	E	30ᵏ		RG
14	10/24	2:20	↑	5+	9"	+	+	+	o	E	30ᵏ		RG
15	10/27	9:24	→	5+	3"	O	+	+	O	E	30ʳ		RG
16	10/27	1:18	→	5+	7"	+	+	+	O	E	30'		RG
17	10/28	10:52	→	5+	6"	+	+	+	O	E	30'		RG
18	10/28	2:38	→	5+	7"	+	+	+	+	E	30ᵗ		RG
19	10/29	10:17	←	5+	5"	+	+	+	+	E	30		RG
20	10/29	2:06	→	5+	5"	+	+	⊕	⊕	E	30		RG
21	10/30	4:34	→	5+	5"	+	+	+	⊕	E	30ᶜ		RG
22	10/30	1:29	←	5+	6"	+	+	+	⊕	E	30'		RG
23	10/31	11:08	→	5+	6"	+	+	+	O	E	30		RG
24	10/31	4:37	↑	5+	6"	+	+	+	+	E	30"		RG
25	11/1	10:56	↑	5+	4"	+	+	+	O	E	30ᵗ		RG
26	11/1	2:47	→	5+	5'	+	+	+	O	E	30ʳ		RG
27	11/2	10:42	→	5+	5"	+	+	+	O	E	30ᵏ		RG
28	11/2	1:37	→	5+	4"	+	+	+	O	E	30'		RG

Figure 8. Page from a protocol illustrating manner of scoring mirror display in squirrel monkeys. Testing twice a day, observer notes magnitude and latency of erection, as well as incidence of the other components of the display—namely, vocalization (*Voc*), thigh-spread (*TS*), scratching (*Scr*), and urination (*Ur*). Arrows in column labeled (*Pos*) indicate direction of monkey's movements in cage during 30-second exposure to a full-length mirror.

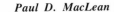
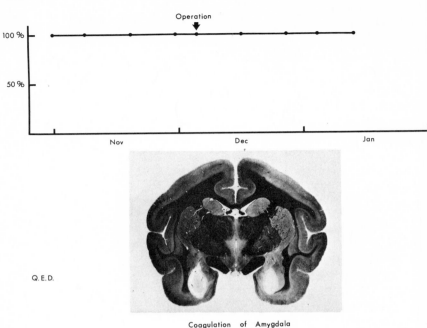

Coagulation of Amygdala

Figure 9. Performance curve of monkey, *Q.E.D.*, illustrating that large bilateral lesions of the amygdala resulted in no change in the animal's perfect display performance. Each point on curve represents an accumulation of 10 trials. Testing began the day after operation.

In experiments on the striatal complex, however, I found that bilateral lesions of the globus pallidus have a profound effect on the monkey's inclination to display (MacLean, 1972, 1973a, 1974b). Fourteen animals were used in this particular series. If the pallidal lesions do not involve the internal capsule, there may be no apparent motor incapacity. Although monkeys with lesions in the anterior or intermediate parts of the globus pallidus may require hand-feeding for the first few days, it should be emphasized that all but 3 of 14 animals maintained or bettered their weight during several weeks or months of testing. When these animals were introduced into our established colony of monkeys, they were able to defend themselves and even overpower the dominant animal. Without a test of the innate display behavior, one might have concluded that these animals were unaffected by the brain lesions.

Figure 10 shows the performance curve of one animal in which bilateral lesions were initially placed in the caudal part of the globus pallidus and contiguous structures. After operation there was a cessation of display for 30 trials and then a recovery of perfect performance. Two additional coagulations shown in Figure 11 were then placed further forward in the globus pallidus. Thereafter there was a persistent failure to display. It should be noted that in this case the lesions encroached on the internal capsule and that there was some motor disability of the hind legs. When the monkey was introduced into our testing colony, however, it was not only successful in resisting attack but also in overpowering the dominant animal. It had been previously shown in other animals that a motor disability of one hind leg resulting from an extensive lesion of the motor cortex did not interfere with the somatic components of the display, including thigh-spreading of the involved leg.

Planimetric measurements were obtained for calculating the

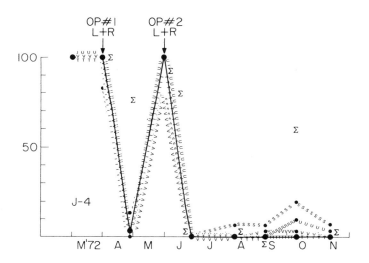

Figure 10. Performance curve of display of Monkey J-4 before and after bilateral lesions were placed in the caudal part of the globus pallidus and then later at a more rostral level. Each large black dot represents an accumulation of 30 trials. The curves shown by letters give percentage incidence of various elements of the display: V, vocalization; T, thigh-spread; U, urination; S, scratch. Σ refers to penile index, a value of the average magnitude of erection whenever it was observed during the course of 30 trials.

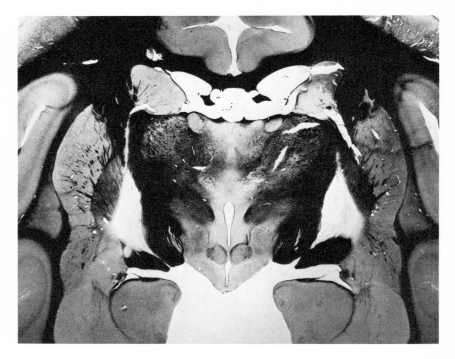

Figure 11. Histological section showing bilateral, symmetrical lesions of the globus pallidus that were made during the second operation on monkey J-4. See caption for Figure 10.

amount of pallidal damage in the 14 cases under consideration. The shaded areas in Figures 12a and 12b represent the amount and locus of the destruction in the left and right globus pallidus in relation to the frontal planes of the brain atlas. The solid black shading indicates a statistically significant difference in postoperative testing. Table I lists both the absolute and percentage values for the amount of destruction.

The analysis revealed that, with destruction of about 16% or more of the globus pallidus, one may expect a highly significant decline in display performance. At the same time it is evident that the amount of destruction is not the only critical factor. It is known that there are rostrocaudal, as well as mediolateral, differences in the projections of the globus pallidus. Figure 12 illustrates that for the combined group of animals the lesions covered most of the rostrocaudal extent of the globus pallidus. It was found that small, complete, bilateral lesions of the posterior globus pallidus had no enduring significant effects on the display.

The projection from the medial segment of the globus pallidus has a different distribution from that of the lateral segment. The analysis showed that bilateral lesions of the lateral segment—the part that projects to the subthalamic nucleus—do not seem to be effective in modifying the display. The most effective lesions appeared to be those involving parts of the globus pallidus where there is a confluence of fibers forming either the ansa or fasciculus lenticularis.

It should be noted that a few animals showed a deterioration of display performance over a period of weeks or months, perhaps indicating that a progressive loss of pallidal neurons or transsynaptic degeneration in other structures may have been significant factors.

In regard to the poorly defined peripallidal collections of gray matter, it is significant that bilateral sublenticular lesions of the nucleus of the ansa peduncularis and of the basal nucleus of Meynert did not affect display performance. It was also found that bilateral interruption of the neighboring stria terminalis and Arnold's (temporal) bundle had no effect on the display.

Pathways for Prosematic Behavior

I am currently investigating the effects of interrupting various outflows from the globus pallidus (MacLean, 1975d). Thus far operations have been performed on 30 animals in which small lesions have been placed in various parts of the subthalamus, hypothalamus, and midbrain. Figure 13 illustrates the histological picture in one animal in which the coagulations involved the confluence of the ventral and dorsal divisions of the ansa lenticularis. This monkey appeared so normal following the operation that there was a question as to whether or not lesions had been made, but it never displayed again during two months of formal testing.

With lesions involving the medial forebrain bundle, a monkey may retain the somatic components of the display but be capable of only partial erection, a finding to be expected on the basis of our earlier observations on the effects of electrically stimulating the medial forebrain bundle (MacLean & Ploog, 1962).

Having briefly mentioned these findings, I wish to emphasize

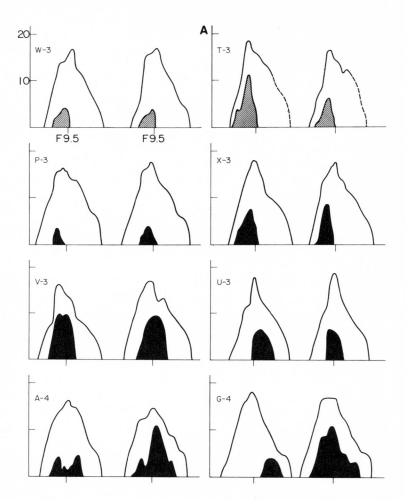

Figures 12A and B. Shaded areas represent amount of destruction of left and right globus pallidus in 14 monkeys. Estimates are based on planimetric measurements. Upper curves show plot of total areal measurements of the globus pallidus made at every half millimeter, rostral and caudal to frontal plane F-9.5 of the stereotaxic atlas (Gergen & MacLean, 1962). Solid black indicates cases in which there was a statistically significant decline in display performance. Monkey T-3 was not formally tested.

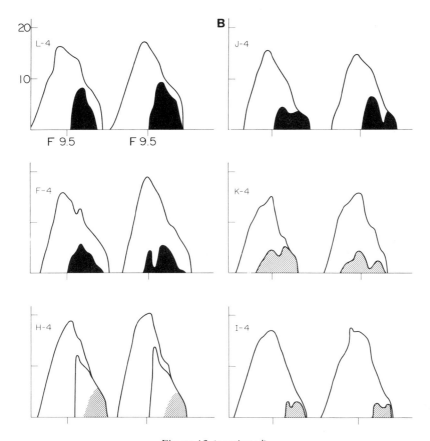

Figure 12 (*continued*)

that the major pathways to and from the reptilian and paleomammalian formations pass through the hypothalamic and subthalamic
regions. If one destroys the majority of these pathways in monkeys,
the animals are greatly incapacitated but, with careful nursing, may
recover the ability to feed themselves and move around. They retain,
of course, the great motor pathways from the neocortex. The striking
thing about these animals is that, although they look like monkeys,
they no longer behave like monkeys. Almost everything that one
would characterize as species-typical simian behavior has disap-
peared. If one were to interpret the experimental findings in the light
of various clinical case materials, one might say that these large

Table I. Volume of Left and Right Globus Pallidus (GP) Destroyed in Respective Animals

Animal[a]	Globus pallidus (total volume, mm)	Amount of destruction			
		Total (% mm³)	L. GP (% mm³)	R. GP (% mm³)	Parts in common (% mm³)
P-3*	144	5 (7)	4 (3)	6 (4)	3 (4)
I-4	141	7(10)	7 (5)	7 (5)	5 (7)
W-3	120	8(10)	9 (5)	8 (5)	7 (9)
H-4	156	8(13)	9 (6)	8 (6)	8(13)
X-3***	135	16(21)	17(12)	14 (9)	10(13)
F-4***	138	16(22)	14(11)	17(11)	10(13)
T-3[b]	127	17(22)	21(15)	12 (7)	10(12)
U-3***	114	19(22)	20(12)	17(10)	2 (2)
K-4	111	21(24)	26(14)	17(10)	13(14)
A-4***	132	23(31)	13 (9)	34(22)	12(16)
J-4***	113	24(27)	25(14)	23(13)	15(17)
L-4***	143	24(35)	20(14)	29(21)	18(26)
G-4***	150	25(37)	9 (7)	39(30)	5 (8)
V-3**	132	32(43)	36(21)	30(22)	19(25)
Mean	132 ± 14	18(23)	16(10)	19(13)	10(13)
		Nonlesioned animals			
H.C. 1	147	—	—	—	—
H.C. 2	151	—	—	—	—

[a] Significant values for mirror display test are as follows: * $p < 0.05$ at termination of experiment; ** $p < 0.025$ at termination of experiment; and *** $p < 0.0005$ at termination of experiment.
[b] Formal testing was not conducted on Monkey T-3.

pathways passing to and from the reptilian and old mammalian formations provide the avenues to the basic personality. Here, certainly, would seem to be essential pathways for the expression of prosematic behavior.

Comment

The experiments that I have described are of special interest because of the demonstration for the first time in a mammal that the striatal complex is involved in genetically constituted, species-typical, prosematic behavior. Since the mirror display also involves

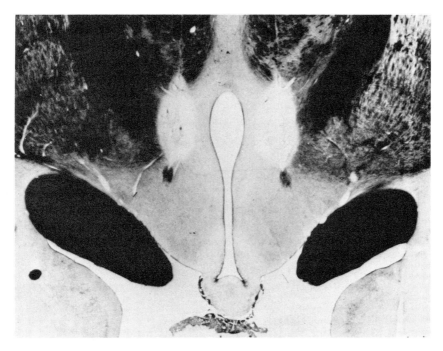

Figure 13. Histological section showing bilateral, symmetrical lesions involving the confluence of the ventral and dorsal divisions of ansa lenticularis. The monkey (P-4) appeared so normal during the first postoperative week that there was a question whether or not lesions had been made. It never displayed, however, during two months of formal testing.

isopraxic factors, the results also indicate that the striatal complex is implicated in natural forms of imitation. In circular language, a species might be defined as a group of animals that has genetically acquired the perfect ability to imitate itself (MacLean, 1975a). The findings that have been presented are not only relevant to species isopraxis but also to what one might call phyletic isopraxis, referring to prototypical patterns of behavior found among animals over a wide range of the phylogenetic scale.

No one would argue that instincts, as generally understood, play a significant role in human behavior. But how should we characterize those proclivities that find expression in the struggle for position and domain; obsessive–compulsive behavior; slavish conformance to old ways of doing things; superstitious actions; obeisance to precedent, as in legal and other matters; devotion to ritual; commemorative, ceremonial re-enactments; resort to treachery and deceit; and imitation?

Many people claim that all human behavior is learned. In their book on social imitation, for example, Miller and Dollard (1941) began by saying, "all human behavior is learned." If so, how does it happen that human beings, with all their intelligence and culturally determined behavior, engage in the basic, ordinary things that animals do?

It is easy to draw parallels between reptilian and mammalian patterns of behavior. Of the many prototypical patterns of behavior enumerated above, I'll mention just two for illustration. Take, for example, the important question of deceptive and predatory behavior, about which almost nothing is known in regard to the underlying brain mechanisms. In the attempted assassination of George Wallace, Arthur Bremer stalked his victim for days at a time and, if he was not around, went for bigger game. Resorting to a kind of reptilian rhetoric, one might ask, "Did Arthur Bremer learn to do this by reading Auffenberg's account (1972) of the predatory and deceptive behavior of the giant Komodo lizard?" These animals, growing up to ten feet in length and weighing 200 pounds, will relentlessly stalk a deer for days at a time or wait in ambush for hours, activities requiring a detailed knowledge of the terrain and a good sense of time. Waiting for just the right moment, the huge lizard will lunge at the deer, cripple it with a slash of the Achilles tendon, and bring it to an agonizing death by ripping out its bowels. The heartless cunning of the Komodo lizard excites curiosity about the possibility that the basic neural circuitry for deception may be built into the striatal complex.

The second example concerns homing, to which I'll return once more at the end of this paper. Homing might be considered as a kind of repetition compulsion that in many instances, because of survival value, seems to be linked up with obeisance to precedent. An example would be the re-enactment behavior of some reptiles in returning to the same place year after year for reproductive purposes. Recent studies have shown that the same situation applies to a number of mammals (Harper, 1970). In a more restricted sense, we recognize the tendency of most animals after exploring and reaching out for food, for mate, or whatever else, to return to a recognized frame of reference. Freud (1922) saw a biological model of human repetition compulsions in the circuitous paths of recapitulation in ontogenetic development and in line with such thinking was led to the conclusion that the goal of all life is death, with the animate returning to the inanimate. It is of tangential interest that recent evidence indicates that the striatal complex contains a large percentage of the brain's opiate receptors (Pert & Snyder, 1973). Couched in neurological terms, one might think of addiction as representing an internalized form of chemical homing.

Question of Integrative Functions

Space does not permit a consideration of the two other main formations of the triune brain, but two brief comments about their functions are required for making a hypothetical point about the adjuvant role of the striatal cortex. The paleomammalian formation is represented by the so-called *limbic system,* that comprises the phylogenetically old cortex of the limbic lobe (see Figure 14) and its related structures within the brain stem (MacLean, 1952). There is both clinical and experimental evidence that the limbic system derives information in terms of emotional feelings that guide behavior required for self-preservation and the preservation of the species. The neomammalian formation includes the neocortex (Figure 14) and structures of the brain stem with which it is primarily connected. It has long been recognized that the neomammalian formation is preeminent in problem solving and other so-called intellective behavior. Figure 15, based on recent anatomical findings, shows schematically connections by which the limbic cortex and neocortex might influence the striatal complex. It has been implicit from the work that I have summarized in the present paper that the striatal complex represents a neural repository for programming behavior based on ancestral learning and ancestral memories. Elsewhere, I have suggested that nature economically utilizes the striatal complex for the re-enactment of currently learned behaviors that have been emotionally conditioned through limbic functions or intellectually mastered by the neomammalian brain (MacLean, 1972).

In the field of literature it is recognized that there are an irreducible number of basic plots and associated emotions. In metaphorical terms one might imagine that the counterpart of the reptilian brain provides the basic plots and actions; that the limbic brain influences

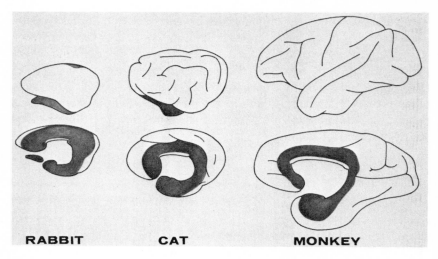

Figure 14. The limbic lobe of Broca (shaded) is found as a common denominator in the brains of all mammals. It contains the greater part of the phylogenetically old cortex, which, together with its related structures in the brain stem, constitutes the so-called *limbic system* (MacLean, 1952). This system represents an inheritance from lower mammals. The neocortex (shown in white), which mushrooms late in evolution, is the hallmark of the neomammalian brain (from colored plate in MacLean, 1973b).

emotionally the development of the plots; that, finally, the neomammalian brain has the capacity to elaborate upon the basic plots and emotions in as many ways as there are authors.

Concluding Remarks

A moment ago I mentioned homing. It is almost two years since Danny Lehrman visited our new laboratory near Poolesville, Maryland, and spoke to us about his work on the ringdove. It was the first and last time I saw him. In retrospect, it would appear that he knew then that he was homing. It was an unusually beaufitul day at the lab; bluebirds were all around, and the sliding glass doors of the library–conference room were open to the meadow. Occasionally during the

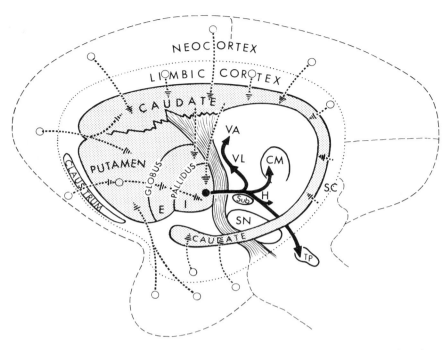

Figure 15. Ringlike configuration of the striatal complex as it would appear in a lateral view. Note how it is surrounded by the limbic lobe. Recent neuroanatomical studies have provided evidence of projections from the greater part of the limbic and neocortex to the corpus striatum (broken lines and arrows) which, in turn, projects to the globus pallidus. Diagram indicates only the outflow from the internal (I) segment of the globus pallidus. Abbreviations: CM, N. centri mediani thalami; E, external segment of globus pallidus; H, area tegmentalis (Forel); SC, Colliculus superior; SN, Substania nigra; Sub, Corpus subthalamicum; TP, N. tegmenti pedunculopontinus; VA, N. ventralis anterior thalami; VL, N. lateralis ventralis thalami.

seminar, Dan would pause when he heard a bird outside and say, "Listen, listen," and then give you the name of the bird. Later we walked down the grassy slope towards the habitats. Again he would stop and say, "Listen, listen." I was not always sure that I could detect what he was hearing.

What you learned from Danny was that when the time for homing comes, you do not wait around wringing your hands. You learn to listen, and keep listening, for all the beautiful things around you!

Summary

The primate forebrain evolves and expands along the lines of three basic patterns characterized as reptilian, paleomammalian, and neomammalian. Radically different in chemistry and structure and in an evolutionary sense countless generations apart, the three formations constitute three brains in one, a *triune* brain. After introductory remarks on the analysis of prosematic (nonverbal) behavior, the central focus of attention is on the hypothesis that the major counterpart of the reptilian forebrain in mammals plays a basic role in complex forms of species-typical, prosematic behavior. Evidence in support of this hypothesis is provided by an extensive investigation of the effects of various brain lesions on the species-typical display behavior of squirrel monkeys (*Saimiri sciureus*). The greater part of the paper is devoted to a summary of the experimental findings and a discussion of the implications.

ACKNOWLEDGMENTS

I wish to express my special appreciation to Mr. Levi Waters, who has helped with the care and postoperative nursing of the animals since the present studies were begun in 1961. I also wish to thank Mr. Robert Gelhard for his invaluable assistance in testing the recent series of monkeys, in the analysis of data, and in the preparation of the figures; Mr. George Creswell and Miss Thalia Klosteridis for histological help; and Mrs. Barbara Coulson for typing the manuscript.

References

Auffenberg, W. Komodo dragons. *Natural History*, 1972, *81*, 52–59.

Bridgman, P. W. *The way things are*. Cambridge, Mass.: Harvard University Press, 1959.

Falck, B., & Hillarp, N. A. On the cellular localization of catecholamines in the brain. *Acta Anatomica*, 1959, *38*, 277–279.

Freud, S. *Beyond the pleasure principle*. Tr. by C. J. M. Hubback. London, Vienna: The International Psycho-Analytical Press, 1922.

Gergen, J. A., & MacLean, P. D. *A stereotaxic atlas of the squirrel monkey's brain (Saimiri sciureus)*. Public Health Service Publication No. 933. Washington, D.C.: U.S. Government Printing Office, 1962.

Greenberg, N. B., Ferguson, J. L., & MacLean, P. D. A neuroethological study of display behavior in lizards. *Society for Neuroscience*, 1976, *2*, 689.

Harper, L. V. Ontogenetic and phylogenetic functions of the parent–offspring relationship in mammals. In D. Lehrman, R. A. Hinde, & E. Shaw (Eds.), *Advances in the study of behavior.* New York: Academic Press, 1970. Pp. 75–117.

Johnston, J. B. The development of the dorsal ventricular ridge in turtles. *Journal of Comparative Neurology,* 1916, *26*, 481–505.

Juorio, A. A., & Vogt, M. Monoamines and their metabolites in the avian brain. *Journal of Physiology*, 1967, *189*, 489–518.

MacLean, P. D. Some psychiatric implications of physiological studies on fronto-temporal portion of limbic system (visceral brain). *Electroencephalography and Clinical Neurophysiology*, 1952, *4*, 407–418.

MacLean, P. D. Contrasting functions of limbic and neocortical systems of the brain and their relevance to psychophysiological aspects of medicine. *American Journal of Medicine*, 1958, *25*, 611–626.

MacLean, P. D. New findings relevant to the evolution of psychosexual functions of the brain. *Journal of Nervous and Mental Disease,* 1962, *135*, 289–301.

MacLean, P. D. Mirror display in the squirrel monkey, Saimiri sciureus. *Science*, 1964, *146*, 950–952.

MacLean, P. D. The brain in relation to empathy and medical education. *Journal of Nervous and Mental Disease, 1967, 144*, 374–382.

MacLean, P. D. The triune brain, emotion, and scientific bias. In F. O. Schmitt (Ed.), *The neurosciences second study program.* New York: The Rockefeller University Press, 1970. Pp. 336–349.

MacLean, P. D. Cerebral evolution and emotional processes: New findings on the striatal complex. *Annals of the New York Academy of Sciences*, 1972, *193*, 137–149.

MacLean, P. D. Effects of pallidal lesions on species-typical display behavior of squirrel monkey. *Federation Proceedings*, 1973a, *32*, 384.

MacLean, P. D. The brain's generation gap: Some human implications. *Zygon; Journal of Religion and Science*, 1973b, *8*, 113–127.

MacLean, P. D. A triune concept of the brain and behavior, Lecture I. Man's reptilian and limbic inheritance; Lecture II. Man's limbic brain and the psychoses; Lecture III. New trends in man's evolution. In T. Boag & D. Campbell (Eds.), *The Hincks memorial lectures.* Toronto: University of Toronto Press, 1973c.

MacLean, P. D. The triune brain. *Medical World News*, Special Supplement on "Psychiatry," October, 1974a, *2*, 55–60.

MacLean, P. D. Brain mechanisms of social behavior. In Society for neuroscience, Third annual meeting, *BIS Conference Report* (36). Los Angeles: Brain Information Service/BRI Publications Office, University of California, 1974b. Pp. 23–26.

MacLean, P. D. The imitative–creative interplay of our three mentalities. In H. Harris (Ed.), *Astride the two cultures: Arthur Koestler at 70.* London: Hutchinson, 1975a. Pp. 187–211.

MacLean, P. D. On the evolution of three mentalities. *Man–Environment Systems*, 1975b, *5*, 213–224.

MacLean, P. D. An ongoing analysis of hippocampal inputs and outputs: Microelectrode and neuroanatomical findings in squirrel monkeys. In R. L. Isaacson & K. H. Pribram (Eds.), *The hippocampus*. Vol. 1. New York: Plenum, 1975c. Pp. 177–211.

MacLean, P. D. Role of pallidal projections in species-typical behavior of squirrel monkey. *Transactions of the American Neurological Association,* 1975d, *100*, 25–28.

MacLean, P. D., & Ploog, D. W. Cerebral representation of penile erection. *Journal of Neurophysiology*, 1962, *25*, 29–55.

Miller, N. & Dollard, J. *Social learning and imitation*. New Haven: Yale University Press, 1941.

Parent, A., & Olivier, A. Comparative histochemical study of the corpus striatum. *Journal für Hirnforschung*, 1970, *12*, 75–81.

Pert, C. B. & Snyder, S. H. Opiate receptor: Demonstration in nervous tissue. *Science*, 1973, *179*, 1011–1014.

Ploog, D. W., & MacLean, P. D. Display of penile erection in squirrel monkey (Saimiri sciureus). *Animal Behaviour*, 1963, *11*, 32–39.

Ploog, D. W., Kopf, S., & Winter, P. Ontogenese des Verhaltens von Totenkopf-Affen (Saimiri sciureus). *Psychologische Forschung*, 1967, *31*, 1–41.

Smith, G. E. A preliminary note on the morphology of the corpus striatum and the origin of the neopallium. *Journal of Anatomy, London*, 1918–19, *53*, 271–291.

Index